T0074449

Buckminster Fuller's World Game and Its Legacy

This book studies R. Buckminster Fuller's World Game and similar world games, past and present.

Proposed by Fuller in 1964 and first played in colleges and universities across North America at a time of growing ecological crisis, the World Game attempted to turn data analysis, systems modelling, scenario building, computer technology, and information design to more egalitarian ends to meet human needs. It challenged players to redistribute finite planetary resources more equitably, to 'make the world work'. Criticised and lauded in equal measure, the World Game has evolved through several formats and continues today in correspondence with debates on planetary stewardship, gamification, data management, and the democratic deficit. This book looks again at how the World Game has been played, focusing on its architecture, design, and gameplay. With hindsight, the World Game might appear naïve, utopian, or technocratic, but we share its problems, if not necessarily its solutions.

Such a study will be of interest to scholars working in art history, design history, game studies, media studies, architecture, and the environmental humanities.

Timothy Stott is Associate Professor in Modern and Contemporary Art History, Department of History of Art and Architecture at Trinity College Dublin, the University of Dublin.

Routledge Focus on Art History and Visual Studies

Routledge Focus on Art History and Visual Studies presents short-form books on varied topics within the fields of art history and visual studies.

Duchamp, Aesthetics, and Capitalism
Julian Jason Haladyn

Post-Conflict Monuments in Bosnia and Herzegovina
Unfinished Histories
Uroš Čvoro

Robert Motherwell, Abstraction, and Philosophy
Robert Hobbs

Jimmie Durham, Europe, and the Art of Relations
Andrea Feeser

World-Forming and Contemporary Art
Jessica Holtaway

The Power and Fluidity of Girlhood in Henry Darger's Art
Leisa Rundquist

Buckminster Fuller's World Game and Its Legacy
Timothy Stott

For more information about this series, please visit: https://www.routledge.com/Routledge-Focus-on-Art-History-and-Visual-Studies/book-series/FOCUSAH

Buckminster Fuller's World Game and Its Legacy

Timothy Stott

Routledge
Taylor & Francis Group

NEW YORK AND LONDON

First published 2022
by Routledge
605 Third Avenue, New York, NY 10158

and by Routledge
2 Park Square, Milton Park, Abingdon, Oxon, OX14 4RN

Routledge is an imprint of the Taylor & Francis Group, an informa business

Library of Congress Cataloging-in-Publication Data
A catalog record for this title has been requested

ISBN: 978-0-367-48390-6 (hbk)
ISBN: 978-1-032-05839-9 (pbk)
ISBN: 978-0-367-48391-3 (ebk)

Typeset in Times New Roman
by codeMantra

Contents

Figures

Preface

'World politics in the Anthropocene cannot be business as usual'.[1]

R. Buckminster Fuller first proposed the World Game in 1964 but had anticipated it for several years. It was first played in colleges and universities across North America from the late 1960s as an attempt to turn a technocratic apparatus of data analysis, systems modelling, scenario building, computer technology, and information design – the stuff of Cold War military strategy – to more egalitarian ends to meet human needs. It challenged players, who were usually not experts, to redistribute finite planetary resources more equitably and, to use Fuller's phrase, to 'make the world work'. The media theorist Gene Youngblood, an early advocate of the World Game, called it 'technoanarchy', a bottom-up, decentralised experiment to repurpose computational and communicational technologies for human benefit. More controversially, Fuller sought to replace politics with 'design science' and the World Game freed the planet's resources from land rights, trade agreements, labour representation, and other local contingencies, just as it assumed US sovereignty over government, infrastructure, and computing.

Unsurprisingly, evaluations of the World Game have been mixed. Some claim that it was a 'viable idea' but politically and economically naïve.[2] Others, more critical, begin from the World Game's unrealised implementation at Expo 67 in Montreal, where Fuller planned to run the World Game in the US Pavilion (discussed in Chapter 1), creating 'a transnational deliberative forum in which world-citizens would use American information technology'.[3] Even if Fuller's plan had gone ahead, the Game's architecture and technocratic approach would have restricted its sought-for deliberative process.[4]

This criticism indicates the determination of the World Game's gameplay and capacity for participation by its design. Everything depended upon how the World Game was set up, especially how its visual technologies designed players' interactions. My proposal in this book is that if we follow this design through the World Game's actual, rather than proposed, architectures, along with similar world gaming projects, we might arrive at an alternative evaluation of the World Game and its legacy. This is not just of historical importance. The World Game's proposals to model planetary 'life support' systems, redistribute resources, and establish a distributed management network foreshadow the needs and dilemmas of current earth system governmentality. Far from replacing politics, and despite Fuller's claims, the World Game was an early attempt to introduce planetary, science-led action and long-term thinking to a non-expert public. Discussions of planetary management and stewardship or 'earth system governance' now shape our politics, as negotiations, conflicts, and alliances become 'climatised'.[5]

This book offers a brief, revised analysis of the World Game and cognate experiments. Although mainly a North American experiment, at least until recently, the World Game always had international ambitions and has evolved across several formats, the latest being an updated World Game Workshop under development at the Schumacher Center for a New Economics where it aligns with projects such as the Green New Deal. It looks again at the migration of ideas and personnel, technologies, game architectures, gameplay, and information design and other techniques of visualisation. With hindsight, the World Game might appear naïve or utopian or technocratic. It was likely all, but not only, those things, and on closer examination, we might find that we share the World Game's problems, if not always its solutions.[6]

Notes

1 Frank Biermann, *Earth System Governance: World Politics in the Anthropocene* (Cambridge: The MIT Press, 2014), 11.
2 Andrew Kirk, *Counterculture Green: The Whole Earth Catalog and American Environmentalism* (Lawrence: University Press of Kansas, 2007), 109.
3 Jonathan Massey, 'Buckminster Fuller's Cybernetic Pastoral: The United States Pavilion at Expo 67', *The Journal of Architecture* 11, no. 4 (2006), 463–464.
4 Felicity D. Scott, *Outlaw Territories: Environments of Insecurity/Architectures of Counterinsurgency* (New York: Zone Books, 2016).
5 Richard Ashley, 'The Eye of Power: The Politics of World Modelling', *International Organization* 37, no. 3 (1983), 495–535. Amy Dalmedico and Matthias Heymann, 'Epistemology and Politics in Earth System Modelling: Historical Perspectives', *Journal of Advances in Modelling Earth Systems* 11 (2019), 1139–1152.
6 Eva Lövbrand, Johannes Stipple, and Bo Wiman, 'Earth System Governmentality: Reflections on Science in the Anthropocene', *Global Environmental Change* 19, no. 1 (2008), 7–13.

Acknowledgements

This book began with research for a paper on the World Game Seminar at the College Art Association annual conference in Los Angeles in February 2018. Thank you to the convenors, Maibritt Borgen and Susan Laxton, and to fellow panellists, Gloria Sutton and Michael Sanchez, for their comments on my paper. Thank you also to Mary Ann Bolger, Larry Busbea, Greg Castillo, Bruce Clarke, Sarah Dry, Johanna Gosse, John Hunter, Simon Sadler, and Greg Watson for many stimulating conversations on world gaming. I have tried to take all your comments and criticisms into account.

Although begun in 2017, much of the research and writing for this book has taken place during the pandemic. I would like to extend a special thanks to those who have made archival materials available and offered their assistance in these most difficult times. These include Sam Cole, Caroline Dagbert at the Centre Canadien d'Architecture/Canadian Centre for Architecture, Yvonne J. Deligato at Binghamton University Special Collections, John Gagnon at Schumacher Center for a New Economics Library, Bradshaw Hovey in the School of Architecture and Planning at University at Buffalo, Nick Lambert, Rob Lansdown, Sarah and George Mallen, Mary Manning at University of Houston Special Collections, and Tim Noakes at Stanford University Special Collections. Thank you also to Stewart Brand, The Estate of R. Buckminster Fuller, John McHale and Magda McHale Archives Foundation, and Yale Center for British Art for granting permission to reproduce some of the illustrations in the book.

Lastly, thank you to my editors, Katie Armstrong and Isabella Vitti, for all their support.

1   Aboard Spaceship Earth

'Spaceship Earth' named a technological object, a political arena, and a matter of environmental concern that, by the end of the 1960s, influenced both United Nations (UN) sustainable development policy and countercultural experiments in ecological redesign and sustainable living, especially in the United States.[1] The World Game attempted to model the life-support systems of this Spaceship, to make them visible, playable, governable, and, for some, optimally functional. 'We are all astronauts now', wrote Fuller in his *Operating Manual for Spaceship Earth*, published in 1968,[2] in command of an 'integrally-designed machine which to be persistently successful must be comprehended and serviced in total'.[3] This made the World Game a 'great world logistics game' for planetary systems,[4] the aim of which was to 'make the total system work more efficiently' and for everyone.[5]

Two major architectural projects were proposed for the World Game, namely the US Pavilion for Expo 67 in Montreal and the World Resources Simulation Center (WRSC) planned for the Edwardsville campus of Southern Illinois University (SIU), envisaged as the first stages of a global network of command centres. Neither project was completed as planned, and the actual architectures of the Game were workshops and seminars, still global in scope but less technologically sophisticated.

On 10 May 1965, economist and systems theorist Kenneth Boulding presented a paper titled 'Earth as a Space Ship' to the National Aeronautics and Space Administration (NASA) Committee on Space Sciences. Humans had to move beyond local, short-term thinking.

> Man must live in the whole system, in which he must recycle his wastes and really face up to the problem of increase in material entropy which his activities create. In a space ship, there are no sewers.[6]

Also, in 1965, US ambassador to the UN, Adlai Stevenson, used the metaphor to refer to a fragile, threatened planet for which the international community was responsible.[7] The following year, Boulding argued that the Earth was not an open system but a closed system, visualised by a 'closed sphere' rather than the 'illimitable plane' of the Mercator projection. This closed sphere required a cyclical, solar-powered 'spaceman economy' to replace the 'cowboy economy' of limitless resource extraction and waste.[8] Spaceship Earth would follow cybernetic principles of feedback, regulation, and efficient management. Conversely, life was reduced to its essential conditions as, faced with a looming environmental crisis, the planet was to be placed on life support.[9] In 1966, the influential economist and journalist Dame Barbara Ward (later Baroness Jackson of Lodsworth) published *Spaceship Earth*, in which she argued that a new planetary responsibility, made operative through progressive taxation, would correct geopolitical inequality. 'Our physical unity has gone far ahead of our moral unity', she wrote. 'Our inability to do anything but live together physically is not matched by any of the institutions that would enable us to live together decently'.[10] Largely due to the advocacy of Ward, care for Spaceship Earth also shaped the ideals of sustainable development, institutionalised through the likes of the International Institute for Environment and Development, an environmental policy and action research organisation founded by Ward in 1973, and reoriented official UN policy on the environment. For the 1972 United Nations Conference on the Human Environment in Stockholm, she co-authored the report *Only One Earth: The Care and Maintenance of a Small Planet* with the American microbiologist René Dubos. At this Conference, which charged itself with outlining, as Ward wrote, 'what should be done to maintain earth as a place suitable for human life not only now, but also for future generations', the US counterculture, mediated by entrepreneur Stewart Brand, sought common purpose with the UN.[11]

The metaphor of Spaceship Earth was a call to environmental action. Its holism struck a chord with the US counterculture, exemplified by the *Whole Earth Catalog* (WEC) motto, 'We can't put it together. It is together'. For example, in March 1970, the WEC included a supplement on the World Game, written by Gene Youngblood based on several articles previously published in the Los Angeles Free Press, and featured on its cover (Figure 1.1) a

Figure 1.1 Cover of World Game supplement, *Whole Earth Catalog*, March 1970, Menlo Park: Portola Institute Inc. Courtesy of Stewart Brand.

photomontage of five young hippies playing volleyball with a photograph of the Earth, above a quote from Fuller:

> I travel around the world a great deal, and everywhere I hear humanity saying, 'We are not against any other human beings;

we feel the world ought to work properly'. Everywhere they say it's our politicians that get us into trouble.[12]

The Earth photograph is that taken by the Applications Technology Satellite 3 (ATS-3) in 1967, used on the cover of the WEC's first edition, published in the autumn of 1968. This photomontage asserts the WEC and its subscribers as defenders and agents of ecological holism and presents the World Game as a version of Earthball, a game conceived by Stewart Brand in 1966 in response to the escalation of the Vietnam War to let players 'understand war by appreciating and experiencing the source of it within themselves'.[13] Earthball then became a symbol of Earth Day in the early 1970s and spawned the New Games Movement, led by Bernie de Koven.

The Biggest System

For Brand and others, such whole earth photographs gave a unified image of Spaceship Earth.[14] The World Game, however, sought to model the Spaceship's planetary systems. For Fuller, although planetary systems might be 'extraordinary' and 'unpredictable' (or 'synergistic', in Fuller's jargon – what we might now call emergent), and not reducible to statistical or probabilistic analysis, they never work 'in ways you can't model'. 'The most important and useful work I've been able to do will be achieving this return to modelability', he declared in a 1971 interview.[15]

The World Game's 'systems approach', which Fuller's research assistant John McHale (see Chapter 4) called 'a new social instrument for complex planning', aimed to define the functional requirements of planetary systems and identify 'within their design various feedback sub-procedures which regulate the system towards the desired optimal end function'.[16] This approach adapted the general systems theory (GST) of the Austrian biologist Ludwig Von Bertalanffy, first outlined in the early 1950s,[17] and combined it with economist John von Neumann's game theory, which studies how strategic interaction between rational economic actors produces intentional and non-intentional outcomes.[18]

In more detail, first, GST provided, in the words of Von Bertalanffy, a 'general science of "wholeness"' to study isomorphisms across open, complex systems and integrate the otherwise disparate fields of biology, robotics, information theory, sociology, economics, and psychology.[19] GST made sense of 'the complexity of the world... in terms of wholes and relationships rather than splitting

[it] into its parts and looking at each in isolation'.[20] In notes for an undated World Game presentation at SIU, Fuller made clear that GST enabled him to model 'Spaceship Earth as a closed system'.[21] In *Operating Manual for Spaceship Earth*, he wrote that GST helped him to 'think in terms of wholes'.[22] Medard Gabel, one of the Seminar players who was employed by SIU from 1969 to work on propagating the World Game, wrote that GST was a

> tool for competent and comprehensive problem-solving [because it demonstrated that] the known behaviour of the whole system and the known behaviour of some of its parts makes possible discovery or true prediction of the remainder of its parts.[23]

This parts/whole correlation made the World Game's strategies 'as comprehensive and correct as is presently possible'.[24] Knowledge of the whole, or what Fuller called the 'biggest system', meant that players could 'automatically avoid leaving out any strategically critical variables'.[25] With this knowledge, too, design scientists – the crew of Spaceship Earth – could learn the 'universe's game' and plan the future. 'If you know something about your game', Fuller declared in interview in 1964, 'you dare to look ahead'.[26] Expo 67 and the WRSC aimed to visualise and operationalise this 'biggest system' in their dome architectures and immersive displays. In short, they aimed to translate GST's 'science of wholeness' into an informational architecture, or what Fuller called 'a total information integrating medium', that was global in ambition.[27]

Second, the World Game aimed to redirect game theory from its application in defence strategy, where it was central to 'computerised world war games and the theory of world economic warfaring'.[28] In many respects, this Cold War strategising was the politics that Fuller sought to abandon (of which more below). Although Fuller frequently cited game theory as one of the World Game's fundamentals, he just as frequently criticised it, as during his presentation to the Joint National Meeting of the American Aeronautical Society and Operations Research Society in Denver in 1969. 'In playing their war games, military establishments consider themselves apolitical', he argued. They make the neo-Malthusian assumption that resource inadequacy necessarily leads to antagonistic politics.[29] The World Game assumed, by contrast, that resource inadequacy was not a given but a product of unequal distribution.

A further criticism was of the game-theoretical understanding of players and their motives. Among clippings collected by Fuller between 1967 and 1968 was a section by the psychologist Robert H. Davis from *Psychological Research in National Defense Today*, a technical report published in June 1967 by the US Army Behavioural Science Research Laboratory, which criticised game theory, especially its zero-sum games, for assuming both that players are 'perfectly rational and motivated by greed' and that cooperation is always a loss rather than a gain. Instead, 'mixed-motive' games introduced uncertainty about the player, 'his value system, his strategy', and allowed that cooperation provides mutual benefit.[30] As noted by Max Ackerman, one of the graduate assistants employed at Central Illinois University for World Game research from 1969, the Game 'goes beyond zero-sum, non-zero-sum, and mixed-motive theory and it may be a new order of game theory: meta-motivational games'.[31] The World Game was a strategy game, but promoted cooperation to answer the basic needs of all aboard Spaceship Earth and thus promoted self-actualisation, or what Abraham Maslow called 'metamotivation'.[32] With their basic needs met through the World Game, the astronauts of Spaceship Earth could devote themselves to more metaphysical or spiritual pursuits.

Total Environments for Planning

Fuller first proposed the World Game in 1964, when asked by the United States Information Agency to design the US Pavilion at the 1967 World Expo in Montreal. The USIA rejected the World Game but agreed to a three-quarter geodesic dome designed by Fuller and Shoji Sadao, whose four interior platforms, designed and curated by Cambridge Seven Associates on the theme of 'Creative America', featured consumer goods and modern art, photographs of Hollywood stars, and a prototype of the Apollo moon lander.[33] Fuller had proposed that from raised platforms visitors would view a suspended miniature earth, which would unfold into a large Dymaxion world map (of which more below) upon which hundreds of lights showed the distribution of global resources such as energy, water, and minerals. Instead, Cambridge Seven's mix of celebrity and consumerism led one local journalist to call the US Pavilion the 'frivolous American bubble'.[34]

Fuller's proposal for the US Pavilion was one of several proposals for a World Game architecture, and not the most ambitious.

Fuller first proposed a World Game facility at the VI International Union of Architects Congress, London, in 1961. The first plans for this facility, drawn by John McHale five years later, show a rectangular space frame truss raised on four seventy-foot-high 'tensegrity columns' with suspended catwalks. At its centre was a large, low relief Dymaxion map to show historical and live data streams. Above the catwalks, from which visitors could view scheduled displays, were personal consoles through which players could request displays, print relevant data sets, focus upon regional problems and trends, ask questions, or test solutions. McHale summarises:

> The whole map complex would be treated as a dynamic display surface capable of showing a comprehensive inventory of the world's raw and organised resources, together with the history and trending patterns of world peoples' movements and needs.[35]

The most ambitious World Game architecture was the $16 million WRSC, planned by Fuller, McHale, and Thomas B. Turner as a 'total information integrating medium' and a 'total world enveloping display' of global resources data.[36] The WRSC consisted of a four-hundred-foot three-quarter dome connected to a shallow one-hundred-and-ninety-five-foot Temcor dome. The latter housed workshops, classrooms, offices, dining areas, and a pentagonal atrium with a large Dymaxion map at its centre, which could be closed or spread flat. The WRSC complex would provide the 'foundation upon which the game is constructed'[37] and offer a 'support data system for World Game/Spaceship Earth activity'.[38] The President of SIU promoted it as an 'environmental early warning system'.[39] For Fuller, the WRSC's 'total environment for planning' would expand awareness of 'possible decision alternatives' to players in much the same way that the Apollo 11 photographs of the Earth had expanded our conception of 'universe'.[40]

Whereas the Temcor dome was largely a workspace, with a display space at its centre, the larger dome aimed at a 'marriage of man/display' (Figure 1.2).[41] In this latter, audience and players would be elevated one-hundred feet, each in a

> 'pilots' contour chair which can be omnidirectionally controlled by a computer – keeping the audience in line with images on dome surface ... and giving them a sense of 'electronic' participation in image generation.[42]

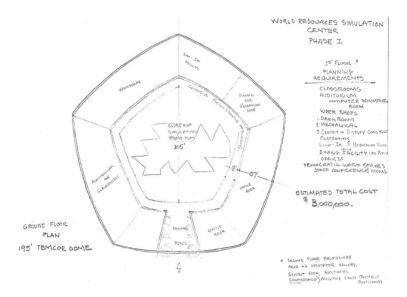

Figure 1.2 R. Buckminster Fuller, Plan for the World Resources Simula-
tion Center, Phase I, c. 1969, Stanford University Collections
M1090, Series 18: Project Files, World Game Subseries 2, 32,
Folder 2: World Game History II, World Game Report (1 of 2).
Courtesy of The Estate of R. Buckminster Fuller.

Such an interface would intensify the sensory experience and expand
the awareness of design scientists, assist their long-term strategies
for resource utilisation, and advance their capacity for planetary
stewardship. The WRSC's 'basic hypothesis', Fuller asserts, is that
decision-making and strategising improve in direct proportion to the
intensity of the visual environment in which they occur.[43] With the
WRSC, as Mark Wigley notes, Fuller pushes architecture further into
the electromagnetic spectrum, as it dissolves into networked image:
'Image has become envelope … Yet again, architecture as the space
of images and the science of architecture as the art of the image'.[44]

   Thomas B. Turner, Director of Research and Development for
the World Game at SIU, wrote in his 'World Game: State-of-the-
Art Report' of December 1969 that the WRSC and the World Game
were necessary to understand our 'dynamic, ever-regenerating environment' and to 'plug in our sensory awareness mechanisms to
the switchboard of "universe", to get in sync with the metabolism
of this spaceship's environment'. As McHale noted, the compound
curvature of such geodesic domes produced 'finite systems', suited

to viewing the world in its entirety.[45] Necessarily, this was a game environment, which would minimise risk to players' decisions.

> [The WRSC] must be perceived, truly, as a "game" environment'. Errors of judgement must not be catastrophic for players or spaceship passengers. ... It must be an environment, in short, which allows him, the decision-maker, to learn ... to hone his awareness of what is possible, of alternative courses of action.[46]

Echoing Fuller, Turner asserts that this game environment had to use the 'visual arts, motion picture, television, theatre, etc. [to] "interface" audience with play so that "audiences" are not just listeners, they are participants'.[47] In this respect, the WRSC further involved Fuller in post-war experiments in immersive 'information machines' for public education in 'worldliness and citizenship', as Nieland writes, just as Charles and Ray Eames had over the previous decade.[48] For the Moscow World's Fair in 1959, he designed a 250-foot-diameter geodesic dome in Sokolniki Park to house the Eames' multi-screen film *Glimpses of the USA*. The dome was described subsequently as an 'information machine'.[49] When Eero Saarinen and Associates constructed a similarly informational architecture for the IBM Pavilion at the New York World's Fair in 1964–1965, which housed *Think*, another multi-screen film composed by the Eames, they included a People Wall to lift five hundred seated visitors into the centre of the visual array.[50] The contour chairs proposed for WRSC were a more interactive version of that Wall.

By 1980, Fuller declared, the WRSC would have the technology necessary to input data rapidly and spontaneously and to project images across the 'entire inner skin surface' of its larger dome. Automated and non-automated pattern recognition systems would identify correlations and trends to planners.[51] As such, the WRSC provided the architectural template for a command centre to pilot and manage Spaceship Earth, in which computers would replace the messiness and fallibility of human decision-making with objective assessments and strategies that would command universal agreement.[52] With time, Fuller believed, it would be possible to extrapolate out players' biases and 'styles' from the assessment of resource scenarios.[53] Seated in their contour chairs, comprehensive design scientists would be the responsible crew of an increasingly automated and streamlined Spaceship Earth. The World Game would become this Spaceship's 'operating software'. Or, as Fuller put it, 'Our World Game will be in effect a World Brain'.[54]

The Temcor dome at Edwardsville was only one stage of a seven-year (1967–1974), multi-phase plan for a World Game Facility. Further stages included the development of an extended twenty-screen 'viewing theatre for montage presentation' of films from the World Game Seminar in NYC and from a Boston-based group led by William Wolf working on 'graphic displays of data reduction and pattern recognition', and a limitless increase in the number of scenarios for playing World Game. 'Since this approaches no upper limit of alternate moves, this process will never really cease'.[55] After parallel activities in Boston, Carbondale, and Edwardsville during Phase II (1970–1974), the aim was to build a permanent facility at Edwardsville 'with specifically modified 4th generation equipment'. None of these proposals were realised. Only a single geodesic dome was built on the Edwardsville campus, designed by Fuller and Sadao in 1971 as a multi-faith religious centre at the request of the SIU Religious Council. It is now the Center for Spirituality & Sustainability.

Beyond Total Environments

At Expo 67, the materials of the Game found a home remote from the 'frivolous American bubble'. A World Trends Exhibit at the Youth Pavilion (designed by architectural firm Ouellet, Reeves & Alain) coincided with a conference on 25 August to mark 'World Architectural Students' Day of the Youth Pavilion', co-organised with the Department of Architecture at McGill University.[56] The schedule for the day included talks by Fuller and Scottish architect Ralph Erskine in the Leacock Auditorium at McGill, followed by a panel discussion at DuPont of Canada auditorium and a late-evening visit to the US Pavilion. The Youth Pavilion itself was situated on La Ronde (Figure 1.3), an extension to the northern end of Île Sainte-Hélène in the Saint Lawrence River and the entertainment sector of Expo 67. The Pavilion's role was education, but was presented as a game for visitors, whom it brought 'apart together in an exceptional situation', to adapt Johann Huizinga's pithy description of play.[57] The Expo 67 Official Guide strived for the youth idiom of the day in describing the atmosphere of La Ronde.

> It's way out, because you're in! And when you play the game, the fellow next to you is not a stranger anymore, because he has come a long way to do the same, and to meet you.[58]

Figure 1.3 Unknown photographer, View of the Youth Pavilion at La Ronde, Expo 67, Montréal, Québec. Architects: Ouellet, Reeves & Alain, 1966–1967, gelatin silver print, 20.3 × 25.4 cm, ARCH256446, Canadian Centre for Architecture, Montréal, Gift of May Cutler.

Visitors to La Ronde discovered materials for the Game alongside a fairground, theatre, musical performances, sports events, art workshops, a puppet festival, and 'happenings'.

The exhibition at the Youth Pavilion belonged to the World Design Science Decade, initiated by Fuller and McHale at SIU in 1965. For two years prior to Expo 67, in Prague, London, Paris, and other major cities in Europe, Africa, Australasia, architecture and design students tested out displays and interfaces for planetary trends in energy production and use, infrastructure, agriculture, resource extraction, and so on. The theme of their first exhibition, beside the Orangerie in the Jardin des Tuileries, Paris, from 2 to 9 July 1965, was 'world literacy regarding world problems', as British architectural students, led by Michael Ben-Eli, sought to identify global problems, their interrelations, and their possible solutions.[59] At the centre of the exhibition was a twenty-foot Miniature Earth upon which were superimposed the world's resources and problem areas. Beside this were panels with photographs showing these same problems accompanied by texts offering design solutions. There was also a Geoscope, which allowed a viewer to enter and see dynamic correlations of trends and resource data. The Geoscope project, begun by McHale and Fuller in 1952, produced a first immersive apparatus to integrate information. Nieland remarks that the Geoscope 'linked resource and data management in [a] futurist quest for a global view of the world *as* information system', a quest continued, of course, in the World Game.[60] Lastly, the Tuileries exhibition included several tensegrity structures made from bamboo, aluminium, and plastic, which were lightweight, portable, and exemplified Fuller's principle of 'ephemeralisation', or doing more with less.

Trends Exhibits such as this were pedagogical tools, prototypes, like the World Game itself, for the *'flexibly dispersed* learning centers for various purposes interlinked to centralised libraries and other major facilities' demanded by McHale for the purposes of world design science.[61] Without these major facilities, however, the World Game existed as a distributed network, where different constituencies would play with future scenarios without end and therefore without 'one best way'. Although the World Game was supposed to integrate scenarios towards an optimal future, it gave no priority to any one scenario. It was, in Fuller's words, an 'aggregate of non-simultaneous and only partially overlapping events'.[62] Each scenario modelled by World Game was, in practice, suboptimal. At the heart of the World Game was an ambivalence. It was an optimal scenario-building game in which nobody wins, or, to extend Fuller's metaphor of mountaineering, a mountain without a peak.

As Reinhold Martin writes, 'By the 1970s ... Fuller was given not so much to projecting ideal futures as to playing games with them'.[63] Martin argues that although Fuller's plans often reflected 'the totalising futurology of the general systems theory that lay behind them', the World Game also 'played with the possibility of reimagining the future as such'. Rather than resemble the system that it sought to replace (that triad of 'Manichean sciences' – operations research, game theory, and cybernetics – criticised by Peter Galison),[64] 'Fuller's project reorients the "system" from within by playing games with the very *idea* of a graspable, collective future'.[65] To do this, the Game displaced politics to cartography (and, I would add, to information design more broadly) to cognitively map the present and extrapolate multiple futures. Its models were provisional and experimental, tested through its data visualisation and its scenarios, which struggled to visualise this 'biggest system'. Likewise, its problem-solving tended to uncover further, often more complex problems.

At stake in the Game, therefore, was

> the possibility of imagining the future as a way out – a way out of the globe, the spaceship, the hotel atrium, the wind-tunnel of progress, and the geodesic dome itself, with its sizzling space frames and its impossible World Games.[66]

The World Game and Its Critics

That the Game remained suboptimal, even despite its ambitions, further complicates its politics. The latter were already ambivalent, both technocratic and anarchic, exemplified by Fuller's claim in 1971 that 'The revolution has come – set on fire from the top'[67] and by Youngblood's description of the 'technoanarchy' that would make it 'possible for society to shape its destiny completely outside the realm of political activity'.[68] Fuller abandoned politics for design science and the World Game forsook the 'political expedient of attempting to reform man and commits man to reforming his environment' through more efficient and egalitarian use of resources aboard Spaceship Earth, as he explained to the Joint National Meeting of the American Aeronautical Society and Operations Research Society in 1969.[69] The World Game would be 'freed from arbitrary political boundaries' and would therefore overcome the inertia of political institutions, Fuller asserted.[70] The design science revolution would lead to an 'incorruptible, true direct planetary democracy with all of humanity franchised and always voting',[71] and cybernetic networks of communications would re-establish 'one-to-one correspondence' between world government and its electorate.[72] 'World Game means to be ready with the answers when the politicians throw their hands up', Fuller declared in an interview for *Life* in 1971.[73]

Fuller's abandonment of politics has drawn much criticism. His appeal to universal humanity and planetary togetherness masked uneven responsibility and consequence of the environmental crisis, just as freeing the world's resources from private or common ownership, land rights, labour costs, and other externalities, assumed US sovereignty over government, infrastructure, and computing technologies needed to establish this 'true direct planetary democracy'. More broadly, design for the purposes of 'planetary housekeeping' (Ward's phrase), more commonly known as sustainable development, has often meant intervention into and management of human populations, settlements, and habitats that eliminates vernacular and endogenous sustainable design practices in the name of development.[74] What is more, the ethnocentric and classist biases of the Spaceship Earth metaphor became evident in concerns that the planet might soon exceed its 'carrying capacity', prompting a moral economy in which only optimum (or optimisable) organisms and systems would justify their place aboard. The 'cabin ecologies' at the heart of the spaceship metaphor, too, brought ecology into line with the Cold War theatre of operations.[75] Nisbet is especially scathing, claiming that World Game transforms ecological responsibility for most into 'simply following the rules predetermined by the top-down principles of global efficiency'.[76]

Many of these criticisms begin from the World Game's proposals for totalising architecture, such as the US Pavilion at Expo 67. Jonathan Massey, for example, writes that the very design of the US Pavilion would have thwarted Fuller's plan to create 'a transnational deliberative forum' even if the World Game had taken place there.[77] Fuller's faith in the objectivity and efficiency of computers, 'legitimated new forms of control', Massey argues.[78] What is more, the Game's 'efficiency-optimising' architecture and 'one best way' approach to resource redistribution would have restricted the deliberative process even as it sought to globalise that process. The Expo 67 proposal appears to exemplify that coupling of 'control in open sites' with the 'quest for "universals of communication"' that once made Gilles Deleuze shudder.[79]

Nevertheless, as noted above, the World Game was suboptimal, despite its grand architectural designs. In addition, its supporters and players often held that its redistributive and egalitarian ambitions could not be reduced to top-down management structures. This becomes more evident once we turn our attention to where and how and with what the World Game was played. We find that in the World Game Seminar, to which we now turn, Spaceship Earth was crewed by amateurs who knew very well their limitations and blind spots, but who muddled along, nonetheless. Indeed, muddling along continues to characterise much of the World Game up to the present.[80]

Notes

1 R. S. Deese, 'The Artifact of Nature: 'Spaceship Earth' and the Dawn of Global Environmentalism', *Endeavour* 33, no. 2 (June 2009), 70.

2 R. Buckminster Fuller, *Operating Manual for Spaceship Earth* (Edwardsville: Southern Illinois University Press, 1968), 46.

3 Fuller, *Operating Manual for Spaceship Earth*, 87.

4 R. Buckminster Fuller, 'How It Came About', in *50 Years of the Design Science Revolution and the World Game* (Carbondale: World Resources Inventory, Southern Illinois University, 1969), 111.

5 R. Buckminster Fuller, 'The World Game', presentation to the Joint National Meeting of the American Aeronautical Society and Operations Research Society, Brown Palace and Cosmopolitan Hotels, Denver, CO. 17–20 June 1969. Paper given on Wednesday 18 June, 12pm to 2.30pm. Stanford University Collections M1090, Series 18: Project Files, World Game Subseries 2, box 25, folder 11, 1.

6 Kenneth E. Boulding, 'Earth as a Space Ship', address to NASA Committee on Space Sciences, Washington State University, 10 May 1965, box 38, Kenneth E. Boulding Papers, University of Colorado at Boulder Libraries.

7 Barbara Ward and René Dubos, *Only One Earth: The Care and Maintenance of a Small Planet* (New York: Norton, 1972), xvii–xviii.

8 Kenneth E. Boulding, 'The Economics of the Coming Spaceship Earth', in *Environmental Quality in a Growing Economy*, ed. H. Jarrett (Baltimore, MD: Johns Hopkins University Press, 1966), 3–14.

9 Sabine Höhler, '"Spaceship Earth": Envisioning Human Habitats in the Environmental Age', *GHI Bulletin* 42 (Spring 2008), 73.

10 Barbara Ward, *Spaceship Earth* (New York: Columbia University Press, 1966), 16.

11 Ward and Dubos, *Only One Earth*, 25.

12 Fuller, quoted on the cover of *Whole Earth Catalog*, March 1970, Menlo Park: Portola Institute Inc.

13 Stewart Brand, 'It Began with World War IV', in *The New Games Book*, ed. A. Fluegelman (New York: Headlands Press, 1976), 8.

14 See Denis Cosgrove, 'Contested Global Visions: *One-World, Whole-Earth*, and the Apollo Space Photographs', *Annals of the Association of American Geographers* 84, no. 2 (June 1994), 290.

15 Barry Farrell, 'The View from the Year 2000', *Life*, 26 February 1971, 53.

16 John McHale, 'World Dwelling', in *The Expendable Reader*, ed. Alex Kitnick (New York: GSAPP Books, 2011), 153.

17 See Von Bertalanffy, 'An Outline of General System Theory,' *British Journal for the Philosophy of Science* I, no. 2 (August 1950), 134–165.

18 R. Buckminster Fuller, keynote address to How To Make The World Work conference at SIU in October 1965, Stanford University Collections M1090, Series 18: Project Files, World Game Subseries 2, box 39, Folder 11: Vision 65 Address, 8.

19 Ludwig Von Bertalanffy, *General Systems Theory: Foundations, Development, Applications* (London: Allen Lane, 1971), 36.

20 M. Ramage and K. Shipp, eds. *Systems Thinkers* (London: Springer, 2009), 1.

21 R. Buckminster Fuller, World Game Presentation at SIU, undated, Stanford University Collections M1090, Series 18: Project Files, World Game Subseries 2, box 24, Folder 14.
22 Fuller, *Operating Manual for Spaceship Earth*, 59.
23 Medard Gabel, 'World Game 'World View'/Frames of Reference Are Composed of...', 5 October 1970, Stanford University Collections M1090, Series 18: Project Files, World Game Subseries 2, box 27, Folder 2, 3.
24 Gabel, 'World Game 'World View'/Frames of Reference Are Composed of...', 3.
25 Fuller, *Operating Manual for Spaceship Earth*, 60.
26 Fuller quoted in Robert Colby Nelson, 'Nature's Extraordinary Order', *Christian Science Monitor*, Tuesday 3 November 1964.
27 Fuller, Vision 65 Address, 3.
28 Fuller, Vision 65 Address, 8.
29 Fuller, 'The World Game', presentation to the Joint National Meeting of the American Aeronautical Society and Operations Research Society.
30 Robert H. Davis, 'International Influence Process: How Relevant Is the Contribution of Psychologists?' in J. E. Ohlaner, ed. *Psychological Research in National Defense Today*, Technical Report S-1, US Army Behavioural Science Research Laboratory, June 1967, 362–363, collected in Stanford University Collections M1090, Series 18: Project Files, World Game Subseries 2, box 33, Folder 7: World Game History 2, Clippings and Reprints, 1967–1968.
31 Letter from Ackerman to Tom Turner, 23 April 1970, Stanford University Collections M1090, Series 18: Project Files, World Game Subseries 2, box 62, Folder 6: World Game Correspondence, Locations, Montreal (1 of 2).
32 A. H. Maslow, 'A Theory of Metamotivation: The Biological Rooting of the Value-Life', *Journal of Humanistic Psychology* 7, no. 2 (1967), 93–127.
33 Jonathan Massey, 'Buckminster Fuller's Cybernetic Pastoral: The United States Pavilion at Expo 67', *The Journal of Architecture* 11, no. 4 (2006), 463.
34 Quoted in Robert Fulford, *Remember Expo: A Pictorial Record* (Toronto: McClelland and Stewart, 1968), 26.
35 John McHale, 'General Description of the Sketch Model', February 1966, Stanford University Collections M1090, Series 18: Project Files, World Game Subseries 2, box 105, Folder 9: Tom Turner Files, World Game Facility.
36 Fuller, 'The World Game', presentation to the Joint National Meeting of the American Aeronautical Society and Operations Research Society.
37 R. Buckminster Fuller, undated presentation at SIU, Stanford University Collections M1090, Series 18: Project Files, World Game Subseries 2, box 24, Folder 14: World Game History, World Game Presentation SIU.
38 Carl G. Nelson, Progress Report on Cooperative Research in World Design, Stanford University Collections M1090, Series 18: Project Files, World Game Subseries 2, box 24, Folder 15: Progress Report (Carl G. Nelson), 1 April 1969.

39 Office of the President (SIU), 'Notes Concerning Plans for the Establishment of the World Resources Computing Center', 1 May 1968, Stanford University Collections M1090, Series 18: Project Files, World Game Subseries 2, box 33, Folder 6: World Game History 2, General Correspondence 1967–1969.

40 Fuller, 'The World Game', presentation to the Joint National Meeting of the American Aeronautical Society and Operations Research Society.

41 Fuller, 'The World Game', presentation to the Joint National Meeting of the American Aeronautical Society and Operations Research Society.

42 R. Buckminster Fuller, Elevation for the World Resources Simulation Center, Phase II, c.1969, Stanford University Collections M1090, Series 18: Project Files, World Game Subseries box 2, 32, Folder 2: World Game History II, World Game Report (1 of 2).

43 R. Buckminster Fuller, 'Planned Implementation of the World Resources Simulation Center, Edwardsville, Illinois', presentation to the Joint National Meeting of the American Aeronautical Society and Operations Research Society, Brown Palace and Cosmopolitan Hotels, Denver, CO. 17–20 June 1969. Paper given on Wednesday 18 June, 12pm to 2.30pm. Stanford University Collections M1090, Series 18: Project Files, World Game Subseries 2, box 25, Folder 1: World Game History, World Resources Simulation Center (2 of 3).

44 Mark Wigley, *Buckminster Fuller Inc.: Architecture in the Age of Radio* (Zurich: Lars Müller, 2015), 267.

45 McHale, 'Buckminster Fuller', in *The Expendable Reader*, ed. Alex Kitnick (New York: GSAPP Books, 2011), 119.

46 Thomas B. Turner, 'World Game: State-of-the-Art Report', December 1969, Stanford University Collections M1090, Series 18: Project Files, World Game Subseries 2, box 25, folder 11 (1 of 3).

47 Turner, 'World Game: State-of-the-Art Report'.

48 Justus Nieland, 'Midcentury Futurisms: Expanded Cinema, Design, and the Modernist Sensorium', *Affirmations: Of the Modern* 2, no. 1 (2014), 56.

49 Beatriz Colomina, 'Enclosed by Images: The Eameses' Multimedia Architecture', *Grey Room* 2 (Winter 2001), 20.

50 See Ben Highmore, 'Machinic Magic: IBM at the 1964–1965 New York World's Fair', *New Formations* 51 (Winter 2003), 128–148.

51 R. Buckminster Fuller, World Resources Simulation Center, Stanford University Collections M1090, Series 18: Project Files, World Game Subseries 2, box 25, Folder 1.

52 From 1951 on, Fuller delivered versions of a lecture on Spaceship Earth. His *Operating Manual for Spaceship Earth* was first published in 1963.

53 Fuller, 'The World Game', presentation to the Joint National Meeting of the American Aeronautical Society and Operations Research Society.

54 Fuller, 'The World Game', presentation to the Joint National Meeting of the American Aeronautical Society and Operations Research Society.

55 Thomas Turner, 'World Game Facility', undated, Stanford University Collections M1090, Series 18: Project Files, World Game Subseries 2, box 105, Folder 9: Tom Turner Files, World Game Facility.

56 Letter from Pierre Bourdon (Coordinator of Youth Pavilion) to John McHale, 5 June 1967. Stanford University Collections M1090, Series 18: Project Files, World Game Subseries 2, box 62, Folder 5: World Game Correspondence, Locations – McGill University (Expo 67).

57 Johann Huizinga, *Homo Ludens: A Study of the Play Element in Culture* (Boston, MA: Beacon Press, 1955), 12.

58 Thérèse Bernard, ed. *Expo 67, Official Guide* (Toronto: Maclean-Hunter Publishing, 1967), 246.

59 John McHale, untitled statement in brochure *Paris 1965*, World Design Science Decade, Stanford University Collections M1090, Series 18: Project Files, World Game Subseries 2, box 62, Folder 8: World Game Correspondence, Locations, Paris (1965).

60 Nieland, 'Midcentury Futurisms', 71–72.

61 McHale, 'World Dwelling', 164.

62 Fuller, audio recording of lecture given at Boston College, 2 May 1970, Stanford University Collections M1090, subseries 5, box 133, 1b, quoted in R. John Williams, 'World Futures', *Critical Inquiry* 42 (Spring 2016), 506.

63 Reinhold Martin, 'Fuller's Futures', in *New Views on R. Buckminster Fuller*, eds. Hsiao-Yun Chu and Roberto G. Trujillo (Stanford, CA: Stanford University Press, 2009), 179.

64 Peter Galison, 'The Ontology of the Enemy: Norbert Wiener and the Cybernetic Vision', *Critical Inquiry* 21, no. 1 (Autumn 1994), 228–266.

65 Martin, 'Fuller's Futures', 179.

66 Martin, 'Fuller's Futures', 187.

67 R. Buckminster Fuller, *No More Secondhand God* (New York: Anchor Books, 1971), 22.

68 Gene Youngblood, untitled, *Los Angeles Free Press*, 19 December 1969. Amended and republished as 'Buckminster Fuller's World Game', *Whole Earth Catalog*, March 1970, 30.

69 Fuller, 'The World Game', presentation to the Joint National Meeting of the American Aeronautical Society and Operations Research Society.

70 R. Buckminster Fuller, *50 Years of the Design Science Revolution and the World Game* (Carbondale: World Resources Inventory, Southern Illinois University, 1969), 118.

71 R. Buckminster Fuller in *Hearings before the Committee on Foreign Relations, United States Senate, Ninety-Fourth Congress, First Session of the United States and the United Nations and the Nomination of Daniel Patrick Moynihan to Be US Representative to the United Nations with the Rank of Ambassador* (Washington, DC: US Government Printing Office, 1975), 235.

72 Fuller in *Hearings before the Committee on Foreign Relations*, 234.

73 Farrell, 'The View from the Year 2000', 50.

74 See Tony Fry, *Design as Politics* (Oxford: Berg, 2010) and Arturo Escobar, *Designs for the Pluriverse: Radical Interdependence, Autonomy, and the Making of Worlds* (Durham, NC: Duke University Press, 2018).

75 On cabin ecologies, see Peder Anker, *From Bauhaus to Ecohouse: A History of Ecological Design* (Baton Rouge: Louisiana State University Press, 2010).
76 James Nisbet, *Ecologies, Environments and Energy Systems in Art of the 1960s and 1970s* (Cambridge, MA: MIT Press, 2014), 77–80.
77 Massey, 'Buckminster Fuller's Cybernetic Pastoral', 463–464.
78 Massey, 'Buckminster Fuller's Cybernetic Pastoral', 478.
79 Gilles Deleuze, in conversation with Antonio Negri, 'Control and Becoming', in *Negotiations, 1972–1990*, trans. M. Joughin (New York: Columbia University Press, 1995), 175. See Rheinhold Martin, *The Organisational Complex: Architecture, Media, and Corporate Space* (Cambridge, MA: MIT Press, 2003), 37.
80 Greg Watson, Director of Policy and Systems Design at the Schumacher Center for a New Economics, in conversation with the author, Monday 18 January 2021.

2 The First World Game Seminar

From 12 June to 31 July 1969 at the New York Studio School of Painting and Sculpture, Buckminster Fuller and Edwin Schlossberg led twenty-six (selected from seventy) graduate students of art, architecture, anthropology, physics, and biology as the first players of the World Game. They had to identify planetary trends and data patterns and build scenarios in which they would redistribute resources (Figure 2.1). On the first two days, Schlossberg recalled, Fuller 'thought aloud' almost fourteen hours, then distributed texts to the players, who, for the next four weeks, studied his writings and projects. From 26 June, the players divided into groups to 'generate preliminary scenarios', which they presented on 29 and 30 July to 'many men and women from all parts of the intellectual and financial and artistic world'.[1]

The World Game modelled planetary systems through the data analytics and information design at its disposal. If Fuller was what Nieland calls a 'data visualisation strategist', so, too, were the Seminar players as they tested Spaceship Earth's 'operating software'.[2] For those involved, the Seminar was a prototype for more sophisticated future versions of the Game.[3] Still, this prototype is important not so much because of the version it might become but because it demonstrates, as Gene Youngblood suggests, the 'vast scope and authority of the World Game even when practiced by amateurs without the optimum technological facilities'.[4] Seminar players used low graphic and informational resolution to interface with planetary systems and, in this, to face what Andrew Pickering calls 'the problematic of getting along performatively with systems that can always surprise us'.[5] As a result, although the World Game was driven by a fantasy of a cybernetic architecture of total information, the cybernetics of the Seminar were characterised by an 'ontology of unknowability'.[6] 'We are in process. It is difficult to see', wrote Schlossberg in his diary entry for 8 July.[7] Seminar players even made a virtue of their ignorance and amateurism. One player

Figure 2.1 Unknown photographer, Photograph of World Game Seminar, Stanford University Collections M1090, Series 18: Project Files, World Game Subseries 2, box 24, folder 1. Courtesy of The Estate of R. Buckminster Fuller.

noted that they were free to work in ways that specialists could not. Another remarked that they were 'amateur in a field where there are no professionals'.[8] 'Ignorance is bliss, isn't it', Schlossberg enthused.[9] Their design science was amateur and partially blind as they took control of Spaceship Earth.

Design and Gameplay of the Seminar

The four steps to the World Game were: (1) Inventory, (2) Trend, (3) Strategy, and (4) Scenario-Building. Players first identified a resource to research, then projected that research into future trends, to provide the world's 'vital statistics'. To identify trends was to bring the inventory 'out of a static state into a dynamic one' and to help players to infer general principles.[10] Once players found principles, they then developed scenarios to provide a synoptic view of planetary developments.[11] Seminar player Medard Gabel described a scenario as,

A logical sequence of events (a strategy) to show how, starting from the present, a future evolutionary condition might evolve step by step; a synergetic synoptic view of as many developments as can be grasped and as may appear relevant to an experimental simulation of a possible reality (the interaction between inventories, trends, and brain).[12]

The principal game rule was cooperation toward more comprehensive and accurate scenarios. The only penalty was disqualification for any player who tried to dominate another or to introduce competition. A team might win a round by solving a problem in the shortest possible time but, in the next round, players would try to improve upon this first, successful scenario, and so on through each subsequent round. This led Schlossberg to declare that the game had no goal. It was iterative and accumulative and could be played without end.[13]

The players' first move was an 'Energy Scenario', 'to design the blood system [of the Earth] so that everybody was fed', Schlossberg wrote.[14] Players plotted the world's energy networks and sought to make them more efficient to meet basic human needs. They calculated the latter in internal metabolics (calories and protein) and external metabolics (kWh), then graphed these from 1965 to 1980 according to current provision to show that most of the world's population will only reach a bare minimum of energy. Their redistribution of off-peak energy, along with more efficient production and transmission, would bring 'everyone on earth to a minimum of 2000 kwh per year by 1980'.[15] This provision, equal to Europe in the late 1960s, was far more than the bare maximum of 1,500 kWh per capita per year needed to make humans 'truly conscious' and able to function collectively as a 'worldwide organism', one player declared.[16] Or, as Schlossberg wrote,

> When we meet the physical needs of all of humanity then the metaphysical [thought and invention] becomes the realm of activity – fear gives way to longing – survival to curiosity – and the spaceship earth begins to care for its crew.[17]

Other students presented on agriculture, food distribution, power infrastructure, mineral and metal resources, levels of industrialisation, and alternative energy sources (wind, solar, geothermal, and tidal). They asked themselves practical questions about needs, materials, and distribution systems, and more speculative questions, such as 'How far ahead can we conceive a future lifestyle?' and

'When is a game a game?'[18] They were troubled, Schlossberg noted, by both the amount of data that was required, and to which they had no access, and with the 'imaginative leap' needed to conceive of Spaceship Earth.

After these first moves, they discussed 'possible synergetic scenarios' that ranged from naïve to utopian to prescient. These included what we now call 'smart' mobile housing, electric cars, biofuels, 'the laser beam transmission of power and information', and 'the possibilities of a world guaranteed annual income'. They also speculated about what humans would do when they were freed from the 'drudgery of having to earn a living'.[19]

For information design, players used a four-foot-high and sixty-foot-long tri-axial grid that showed energy needs, population, population density, access to water, and 'vital statistics' of Earth's inhabitants and two ten-by-fifteen foot Dymaxion maps upon which were placed acetate overlays to show global resources, networks, and infrastructures. Mimicking the jargon of Fuller, Gabel described these as,

> The graphical, functional and mathematical orderings and simplifications of the omni-complicated and interrelated processes of the World. The conceptual simplifications of "reality" into the vectors of an interacting process which can be dealt with on a scientific basis.[20]

This information design was fundamental to the gameplay of these amateurs 'in a field where there are no professionals'. Players used these maps and grids as what Johanna Drucker calls 'knowledge generators', or

> Graphical forms that support combinatoric calculation. Their spatial organisation may be static or mobile, but their spatial features allow their components to be combined in a multiplicity of ways. They make use of position, sequence, order, and comparison across aligned fields as fundamental spatial properties.[21]

Both Dymaxion map and grid were generative and dynamic. They provided players with working models of planetary systems. How they did this I now discuss in more detail, beginning with the Dymaxion projection.

A Faithful Background

Martin argues that the Game displaced politics to cartography. Certainly, at its centre was the Dymaxion projection conceived by Fuller, a world map first published in *Life* in March 1943 as a cuboctahedron, with eight triangular faces and six square faces. Patented in 1946, a decade later, it was republished as an icosahedron, which was the version used by Game players. According to *Life*, this projection satisfied the need 'for a visually correct picture of the earth' because it distributed distortions of scale, direction, and shape evenly across its surface.[22] Distortions 'are nowhere extreme', *Life* enthused.[23] Readers could cut out and recombine the map's fourteen faces or could rotate the entire map through 360 degrees. It also allowed for a topological transformation from solid to plane without distortion. The Dymaxion projection belongs to a class of 'myriahedral' projections that are almost conformal and conserve areas, but which have many interruptions across their surface.[24] Their advantage is to reduce distortion and to allow for three-dimensional and, when unfolded, two-dimensional representation. For Fuller, the minimal distortions of the Dymaxion projection provided players with a 'faithful background upon which to show data'.[25]

Following Denis Wood and John Fels, we might consider the propositional logic of this 'faithful background'. Like all maps, the Dymaxion projection was a vehicle for 'the creation and conveying of authority about, and ultimately over, territory', whose authority is 'the social manifestation of what the map presents as its "intrinsic" and "incontrovertible" factuality'.[26] Evidently, Fuller believed the factuality of the Dymaxion projection to be incontrovertible. This is what made the map a 'faithful background'. Likewise, he and the players assumed the factuality of the data presented upon this background, which gave the World Game authority. The Dymaxion projection presented a world unified as a game board (or theatre of operations) and as a closed system. Wood and Fels argue that maps are 'significant players ... in the world of action'.[27]

> Maps appear as players in a complicated social game defining the relationship of our species to the rest of existence. Pretending to be no more than scorekeepers, maps stand revealed as more like the ball, the very medium through which the game's moves are made.[28]

For *Life*, this projection followed in the footsteps of Harold Mac-kinder's world map of 1904, which showed the so-called 'natural seats of power'.[29] Here, the world of geopolitics pivoted around control of Eurasia, or what Mackinder later called, writing after the First World War, the Heartland or 'World-Island'.[30] As noted, *Life* updated this geopolitical network, complete with its imperial assumptions, for American readers during the Second World War. Recombining its faces would show different geopolitical perspectives and 'animate the facts of geography'. These facts betrayed, unsurprisingly, a bias toward the United States and its allies. Readers learnt of the accomplishments of the British Empire, the 'ruthless logic' of 'Jap imperialism', the German advance into Mackinder's Heartland.[31]

The projection's potential exceeded *Life*'s wartime instructions, of course. Boulding had criticised the Mercator projection for its 'illimitable plane' and the 'cowboy economy' it supported, which treated the world as an inexhaustible resource. By contrast, Fuller's projection would visualise the world as an integrated system not for the purposes of geopolitics but to redistribute resources that were bounded, integrated, and exhaustible. What is more, in anticipation of the now-current ecological metaphor of a blue planet, the Dymaxion projection showed a unified ocean. *Life* declared that the Dymaxion projection showed the world as 'one continent', as the sinuses were placed through the oceans to maintain continuity of land areas. This was a principal feature, explained in detail in the patent (Figure 2.2).[32] Rather than a continent, however, Fuller would later call it a 'one-world island in a one-world ocean'.[33] In truth, it is more of an archipelago than an island.

Wood and Fels add that nature mapped as an integrated system, such as the 'SKY OCEAN WORLD' of the Dymaxion projection, is nature that is known, the object of science.

> Not simply something to be cared for or feared, admired or cuddled or corralled or catalogued, but something truly apart that has to be taken apart, and then put back together in our own terms, to be truly understood … in other words, truly possessed. The epistemological pretence is overwhelming.[34]

Certainly, this was how the Dymaxion projection was presented in *Life*, whether as polyhedral globe or plane. In the World Resources Simulation Center (WRSC) and Expo 67, it was to

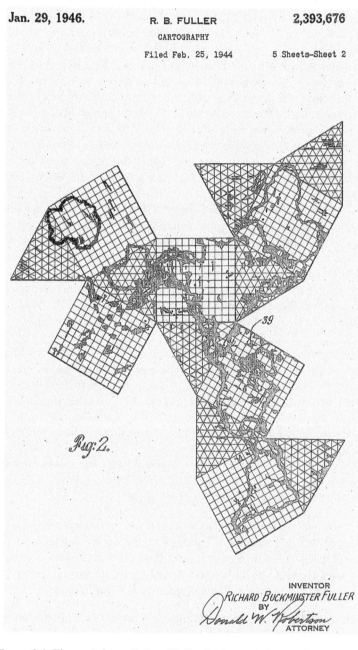

Jan. 29, 1946.   R. B. FULLER   2,393,676

CARTOGRAPHY

Filed Feb. 25, 1944   5 Sheets–Sheet 2

Figure 2.2 Figure 2 from Fuller, R. Buckminster cartography, US patent 23939676, filed 25 February 1944, issued 29 January 1946. Courtesy of The Estate of R. Buckminster Fuller.

perform in both these modalities, suspended then lowered and laid flat. Its 'epistemological pretence' is clear: the planet is knowable and available as a game board and arena of voluntary, global action. In more detail, however, as polyhedral globe, it offers 'an object of contemplation, detached from the domain of lived experience'.[35] With a globe, argues Tim Ingold, 'To know the world ... is a matter not of sensory attunement, but of cognitive reconstruction'.[36] This latter would continue with the Dymaxion projection as plane laid out on the floor of the WRSC. Yet, the 'compound curvature' of the latter's dome would have functioned as both globe and sphere, in which players would play out detachment and engagement. In the Seminar, players viewed the projection only upon two vertical planes, which they constructed as working models. During their scenario-building through the mapping of planetary systems on acetate overlays, the two Dymaxion maps therefore performed more as sections of the interior of a sphere. Did this change their 'epistemological pretence'? It is difficult to say. Certainly, in this orientation, the Dymaxion clearly shows Fuller's 'one-world island' as more of an archipelago that required dynamic integration through the player's interactions with it. This 'biggest system' could not be assumed simply as an already-integrated background. What is more, this one-world archipelago was surrounded by, to borrow a phrase from Bruno Latour, 'a vast ocean of uncertainties', with which players must interact, collaborate, build approximate scenarios, and negotiate solutions.[37] The epistemological pretence might be total knowledge, but in play, uncertainty enters into the World Game through players' interactions with its visual technologies, such as the Dymaxion projection, through the scalar inflexibility of this latter, and the diverse inferences players might make from these interactions. Fuller's attempt to correlate intention and representation through prefacing the World Game with his talks and publications cannot erase what appears to be the constitutive role of uncertainty in its visual technologies.[38]

Data from a Vast Machine

Another source of uncertainty in the World Game is its use of data. In addition to the acetate overlays placed upon a 'faithful background', Seminar players produced what Schlossberg called a 'triangulated interactive grid' to show correlations between data sets. This provided a 'working storehouse for information' on energy needs, population, population density, access to water, and so on, divided into 120 sections (right diagonals) and 23 'major areas' (left diagonals).[39] Gabel explains in more detail:

> This chart was a triangular grid on which one of the three axes were the 22 major geographical areas of the world and their individual countries. The second axis consisted of, in five year increments from 1965 to [2000], figures in population, population density, calorie and protein intake, total kwh, metric tons of coal equivalents and energy slaves. The last axis could indicate up to 20 possible world-trends for each area and country. We use thirteen: fossil fuel potential, life expectancy, mortality rate, arable land, housing, amounts of copper, aluminium and steel, food literacy, reinvestable time and hydropower.[40]

The players used this grid as 'an object to think with', a computational device for building scenarios.[41]

Returning to Drucker, we might argue that the data shown in this grid 'are *capta*, taken not given, constructed as an interpretation of the phenomenal world, not inherent in it'.[42] The grid depended upon already parameterised data sets, themselves the product of an 'infrastructural globalism' discussed by Paul N. Edwards in his history of how meteorology and climatology constructed an extensive infrastructure, a 'vast machine' to collect, model, and disseminate environmental data.[43] The World Game has its roots in this and other infrastructural globalisms, which precede the whole-earth environmentalism of the late 1960s. Seminar players collected data from a vast data-generating machine that produced, for example, Reports for the UN Statistics Commission and the New York Times Encyclopaedia Almanac, and attempted to make this data accessible and operable. However, the World Game includes no opportunity for players to reflect upon how this data is collected or processed, its accuracy, its relevance to definitions of human needs, or to what extent the design parameters of the Game predetermined the problems in response to which this data would be mobilised.[44]

The triaxial grid allows for combinatoric activity as players chart variables against each other. Retrieving and cross-referencing data from this grid was iterative and accumulative, which meant that, once again, players correlated systems variables without implying a 'biggest system'. Gabel asserted that in a closed system all variables could be controlled,[45] but the open structure of the grid could not model a closed system because it could not include all those 'strategically critical variables' of which, Fuller and other Game advocates believed, automated computation would give a full account. As a result, the grid modelled open systems where the critical character of variables could only be ascertained once they were correlated with other variables. Added to the limitations, omissions, and biases of the players' data sets, players' use and representation of data through information design further introduces uncertainty and complexity. Crucially, the Seminar players seemed to know this.

Playing with Black Boxes

The Game necessarily reduced complex systems to models of manageable complexity, wherein situational possibilities (trends and scenarios) did not exceed those projects available to players by voluntary fiat. Otherwise, there could have been no play, according to the model of play proposed by psychologist Mihaly Csikszentmihalyi and sociologist Stith Bennett only two years after the Seminar.[46] The players knew the limits of their information, their models, and their scenarios. The triangulated interactive grid and the accumulated acetate overlays show an iterative and complex systems approach, which suggests that although the Seminar players were encouraged to imagine themselves as the crew of Spaceship Earth and often deferred to a future environment of total, real-time information, and computational power, they nonetheless flew in twilight, as it were, feeling their way from one Black Box to the next. Rather than a failure of the Game, this returned players to the messy world of politics, education, and dissemination, and to a non-holistic modelling of planetary systems.

Their blindness cannot be reduced to a simple lack of computational power, as the World Game continued despite this lack, suggesting that the latter might be structural, necessary to ongoing prototyping. Felicity Scott recounts Gene Youngblood's journey to a World Game demonstration in 1970. When flying with Fuller and Michael Binelli (a geodesic dome engineer) in a light aircraft from St. Louis to Carbondale, Youngblood realised at some point that they were flying blind. 'I was thinking how appropriate it was to be with Bucky this way, trusting our lives to the very design integrity that he has spent a lifetime defending as humanity's only hope for success'. In this situation, flying blind celebrated computational assistance and reliability.[47] During the Seminar, however, players flew blind because of their models' limitations and the character of their visualisation strategies, a blindness that would have been eased only partly by greater computational power.

Again, it seems that uncertainty and complexity were structural and productive rather than incidental, because still the Seminar players modelled and played. In doing so, they made, to borrow from Bruno Latour, a 'passage through connections' without assuming a Whole or a 'biggest system'. They performed 'connectivity without holism'.[48] This is significant because, if the Earth System itself (what James Lovelock famously called Gaia) is 'anti-systemic', in the sense that there is no superior level above and beyond its parts, then the metaphor of a spaceship, with its pilot's chair, mission control, and design scientists, fails.[49] Instead, players faced the problem of how to get along with complex planetary systems that self-regulate as they co-evolve without 'foresight, planning or teleology'.[50] In such a case, the Game would be no cybernetic apparatus worthy of name, it seems, as there would be no 'tiller of environmental control'.[51] Yet, as noted above, perhaps there is another cybernetics in operation here, with its Black Boxes and 'ontology of unknowability', a more experimental cybernetics. Although they did not use the term, and the Seminar coincided with the Apollo 11 mission (16–24 July), which, in Schlossberg's view, enabled humans to 'see the earth for the first time as a spaceship' (21 July was marked as MOON DAY in his diary), their play was often sub-lunar.[52]

The concept of a Black Box, which arose to describe problems in electrical engineering, extends to any systemic entity that performs some task by means largely obscure to an observer. To paraphrase Ross Ashby, when faced with a Black Box, the question 'What is in it?' becomes redundant. Other questions are more pressing, such as 'How should an observer proceed?' 'Which elements of the Box are discoverable and which are not?' 'What methods can be used to investigate the Box?'[53] Perhaps the only way to engage unpredictable Black Boxes, what mathematician René Thom called their 'catastrophe', is to play with them.[54]

Seminar players admitted that many of their models were reductive and lamented their lack of total information about the systems that they researched. Yet, they accepted their ignorance and played on. They planned to run simulations to see how different systems would respond to their moves, but again, were reluctant to do so because they lacked information. This uncoupled the Game from what McHale called Fuller's 'fanatical belief in complete over-all pre-planning' and made it into a distributed pedagogical tool.[55]

What did the World Game teach? For some, it provided a 'curriculum in environmental studies',[56] but its understanding of ecology and environment turned upon its design and architecture. On the one hand, once again, it aspired to a totalising view. The whole of Spaceship Earth would be visible from the 'conning tower' of the World Game, as one Seminar player put it in a presentation at the International Design Conference at Aspen in 1970. The Game provided an 'aerial view of ecology'. Players were like foresters looking from a tower to see 'what is occurring in the forest'.[57] The WRSC's 'total world enveloping display' supported this planetary metaphysics as it elevated the design scientist in his 'pilots' contour chair' to the status of a sovereign decision maker. It is, of course, a poor ecologist, or a poor student of environmental studies, who investigates the forest ecosystem only from above the canopy. The grid, as noted, modelled systems differently. It offered only accumulated parts, and so returned players to the 'forest floor' (as did the accumulative and distributed nature of the Game's learning centres).

If we do not assume a 'spaceship', and therefore no pilot's chair, no mission control, no 'total world enveloping display' that could guide long-range planning, we are left with the playing of the Game itself, where players felt their way along a 'passage through connections'. In this, the experience and activity of the Seminar players might still offer us some lessons. Their play offers a different modelling of earth systems to the mainstream environmentalism in the United States in the late 1960s and 1970s, which tended to treat the earth as a closed, self-regulating system. The players knew the limits of their scenarios. They also knew that they were amateur pilots. During the public demonstration, one audience member remarked that, however much the players overlooked these political conditions, World Game would have to be implemented. 'How can this be done without politics?' he asked.[58] Gabel replied that the players were playing a 'mountain-climbing game' to show that they could get everybody to the top as rapidly as possible. In his diary, Schlossberg recalls: 'we answer that we were not planning implementation but only exploring whether it is feasible to consider that man can be provided with physical support by the year 2000'.[59] Showing that this was possible, the players believed, would prompt action without waiting for crisis.

After the Seminar

Following the Seminar, many players engaged in the prosaic politics of disseminating and promoting the World Game. Several joined Fuller at Southern Illinois University (SIU) to organise World Game workshops, first in the Department of Design at SIU, 29 June to 21 August 1970, then in Miami from 15 September to 22 December 1970, and

San Francisco from 5 October to 19 December 1970. Other art and design colleges across the United States and Canada offered versions of the Seminar. A starter pack, the World Game Package, was compiled and distributed, which included several Dymaxion maps, a cassette of a Denver radio interview with Fuller, and a Lift-Off Manual of texts by Fuller, Schlossberg, and Youngblood, all for $35, 'a minimum cost for maximum benefit'. A more expensive version, costing $1,000, included films, books, slides, and large-scale Dymaxion maps.[60] There was talk of a board game. By March 1970, there were World Game extension groups in New York, Montreal, New Concord, Los Angeles, and Bridgeport. That summer, at SIU, an eight-week World Game Workshop was scheduled. Tasks included definition of critical variables, model building and testing, game building and refining, with evaluation and debriefing to conclude.[61]

Twice, in December 1970 and March 1971, Michael Ben-Eli, Schlossberg, and Winsey led the seminar 'The World Game for Government Executives' as part of the Special Program of the Graduate School of the US Department of Agriculture in the National Press Building in Washington D.C.[62] They presented the World Game as a 'new educational simulation system' that would provide government executives with 'a complete reality of what we mean by "World", of what problems that world faces and what strategies might solve them. The brochure quotes at length from Fuller's dismissal of the Malthusian doctrine and the militarised geo-politics that derived from it.[63] One of the Seminar's objectives was that participants would 'escape from the constriction of specialised knowledge and look at the world as a spaceship with identifiable and worldwide characteristics'. Winsey ran the second seminar three months later. Participants included Manpower Analysts, an Aerospace Technologist, the Special Assistant to the Director of National Aeronautics and Space Administration (NASA), and a Pastor from Baltimore.[64] In May 1972, she wrote to Fuller requesting his support for a World Game handbook.[65]

Winsey organised an undergraduate course Dynamics of Change at Pace College, based on the World Game (Figure 2.3). She describes the course thus:

> Pace students explore the interdependence of natural, physical, and human resources on the global scale. Perceived as a game, each "player" learns to integrate his own experiences and research with those of others.[66]

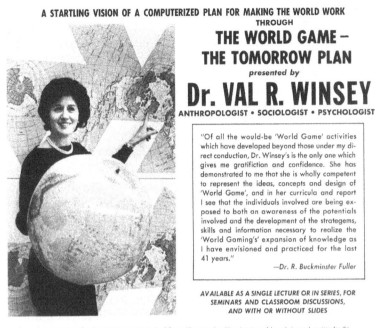

Figure 2.3 Flyer for The World Game-The Tomorrow Plan by Val Winsey, Stanford University Collections M1090, Series 18: Project Files, World Game Subseries 2, box 65, Folder 3, Locations - Pace University (Val Winsey) [2 of 2]. Courtesy of The Estate of R. Buckminster Fuller.

Players had to research global trends, resources, networks, and infrastructures, as they had in the World Game Seminar. Scenarios were then envisaged using acetate overlays on Dymaxion maps. Twenty-five students worked in a dedicated room in Pace College's 150 Nassau Street in Lower Manhattan. With Dymaxion maps on one wall, the room also contained atlases, projectors, and an inflated plastic globe. As with the World Game, politics was denied.

The one World Game requirement is that students believe that local, national and international politics, boundaries, and borders no longer exist.[67]

Despite this, Winsey notes that students of the course also acquired an interdisciplinary education that is otherwise unavailable within academia. She then began a tour of lectures and seminars advocating the World Game authorised by Fuller and managed by the speakers' bureau W. Colston Leigh, Inc.

Research done during this course led a student named Spencer Marks to the 'Co-Op City Community Study'. Based on the site of a former amusement park in north Bronx, NY, the Study took the World Game as a model for a cooperative and proposed a prototype environment, Integron, 'specifically created for synergistic group work on large scale ecological problems'. The World Game had shown the value of non-zero-sum games: 'all players share the same fate. They are synergetic games because the sum total of wins and losses is either greater or less than zero'. The Integron Environment, conceived by James Tackaberry McCay, and created by Quebecois architect Victor Prus, offered

> a broad array of facilities for innovation and re-creation including operational amphitheatre with multi-dimensional information displays, computer area, television studios and video-tape feedback equipment, environment design laboratories, libraries, living accommodations, dining and lounge areas, physiotherapy, swimming and play areas, and maintenance facilities. The Integron is a model for larger, improved comprehensive environments for communities, hospitals, schools, offices and research in the future.[68]

Another early proposal to expand the World Game was made by Phillip H. Lawyer and Richard Meyer in late summer 1970, Research Assistants at SIU Free School at Carbondale, Illinois, working on a project titled 'Spaceship Earth Exploration by Design Science (The World Game)'. They suggested to Dr Marty Groder, Warden of Marion Federal Prison, seventeen miles east of Carbondale, a 'transactional analysis encounter group with the inmates'. The World Game could be used for penal reform, 'applied as a tool in the study of correction', they argued.[69] Transactional analysis had been developed by psychotherapist Eric Berne in several works through the 1960s and 1970s, including *Games People Play: The Psychology of Human Relationships*, published in 1964.[70] The SIU Free School was sponsored by the SIU students through their activities programme, with facilities provided by the University. As with Free Universities worldwide, there were no entrance requirements and no tuition fees. It included courses on auto care, visual media, photography (it had a dark room on campus), arts and crafts, 'spirituality of macrobiotic cooking', and guitar.

However ambitious these latter proposals might be, they show that within a year of the Seminar the World Game had a life beyond Fuller as a distributed pedagogical and research tool. It had a life, that is, within an evolving family of prototypes that broadly shared its ambitions. It is to one of these prototypes, an 'ecogame', that I turn in the next chapter, before returning to pick up the thread of the World Game's development in the final chapter.

Notes

1 Edwin Schlossberg, World Game Diary, Stanford University Collections M1090, Series 18: Project Files, World Game Subseries 2, box 39, Folder 2, 19.
2 Justus Nieland, 'Midcentury Futurisms: Expanded Cinema, Design, and the Modernist Sensorium', *Affirmations: Of the Modern* 2, no. 1 (2014), 71.
3 Gene Youngblood, 'Buckminster Fuller's World Game', *Whole Earth Catalog*, March 1970, 30.
4 Youngblood, 'Buckminster Fuller's World Game', 30.
5 Andrew Pickering, *The Cybernetic Brain: Sketches of Another Future* (Chicago, IL: Chicago University Press, 2010), 23.
6 Pickering, *The Cybernetic Brain*, 23.
7 Schlossberg, World Game Diary, 12.
8 Unidentified student in *The World Game: Playing World Game (Getting power to the people)*.
9 Schlossberg, World Game Diary, 12, 14.
10 Fuller, World Game Presentation at SIU (undated), Stanford University Collections M1090, Series 18: Project Files, World Game Subseries 2, box 24, Folder 14.
11 Medard Gabel, 'World Game 'World View'/Frames of Reference Are Composed of...', 5 October 1970. Stanford University Collections M1090, Series 18: Project Files, World Game Subseries 2, box 27, Folder 2, 3.
12 Gabel, 'World Game 'World View'/Frames of Reference Are Composed of...', 5.
13 Schlossberg, World Game Diary, 9.
14 Schlossberg, World Game Diary, 4.
15 Medard Gabel, 'Buckminster Fuller's World Game', *Whole Earth Catalog*, March 1970, 31.
16 Unnamed student in *The World Game: Playing World Game (A hundred million horses going nowhere)*.
17 Schlossberg, World Game Diary, 12.
18 Gabel, 'Buckminster Fuller's World Game', 31.
19 Gabel, 'Buckminster Fuller's World Game', 32.
20 Gabel, 'World Game 'World View'/Frames of Reference Are Composed of...', 8.

21 Johanna Drucker, *Graphesis: Visual Forms of Knowledge Production* (Cambridge, MA: Harvard University Press, 2014), 105.

22 Anon. 'R. Buckminster Fuller's Dymaxion World', *Life*, 1 March 1943, 42.

23 Anon. 'R. Buckminster Fuller's Dymaxion World', 42.

24 Jarke J. van Wijk, 'Unfolding the Earth: Myriahedral Projections', *The Cartographic Journal* 45, no. 1 (2008), 32–42.

25 Fuller in Herbert Matter, *The World Game: The Structure of Nature (I'm going to take you to breakfast yesterday morning)*, filmed at 1969 NY Studio School, World Game Seminar, Saturn Pictures, 23–30 June 1969. 60 min., b & w, 16 mm. Stanford University Collections M1090, Series 17: Subseries 7, 73a.

26 Denis Wood and John Fels, *The Natures of Maps: Cartographic Constructions of the Natural World* (Chicago, IL: University of Chicago Press, 2008), 7.

27 Wood and Fels, *The Natures of Maps*, 195.

28 Wood and Fels, *The Natures of Maps*, 21.

29 Harold Mackinder, 'The Geographical Pivot of History', *The Geographical Journal* XXIII, no. 4 (April 1094), 435.

30 Harold Mackinder, *Democratic Ideals and Reality: A Study in the Politics of Reconstruction* (London: Constable & Co, 1919).

31 Anon. 'R. Buckminster Fuller's Dymaxion World', 42.

32 R. Buckminster Fuller, cartography, US patent 23939676, filed 25 February 1944, issued 29 January 1946.

33 R. Buckminster Fuller, typescript for 'World Game: How It Came About', 21 April 1968, Stanford University Collections M1090, Series 18: Project Files, World Game Subseries 2, box 27, Folder 2: World Game History, General (2 of 2), 5.

34 Wood and Fels, *The Natures of Maps*, 166.

35 Tim Ingold, 'Globes and Spheres: The Topology of Environmentalism', in *The Perception of the Environment* (London: Routledge, 2000), 209.

36 Ingold, 'Globes and Spheres', 213.

37 Bruno Latour, *Re-Assembling the Social: An Introduction to Actor-Network-Theory* (Oxford: Oxford University Press, 2005), 245.

38 Here, I follow the argument of Matthijs Kouw, Christoph van den Heuvel, and Andrea Scharnhorst, 'Exploring Uncertainty in Knowledge Representations: Classifications, Simulations, and Models of the World', in *Virtual Knowledge. Experimenting in the Humanities and Social Sciences*, eds. P. Wouters, A. Beaulieu, A. Scharnhorst, and S. Wyatt (Cambridge, MA: MIT Press, 2012), 105–110.

39 Schlossberg, World Game Diary, 14.

40 Gabel, 'Buckminster Fuller's World Game', 31.

41 John Bender and Michael Marriman, *The Culture of Diagram* (Stanford, CA: Stanford University Press, 2010), 7.

42 Drucker, *Graphesis*, 128.

43 Paul N. Edwards, *A Vast Machine: Computer Models, Climate Data and the Politics of Global Warming* (Cambridge, MA: MIT Press, 2010).

44 Kouw, van den Heuvel, and Scharnhorst, 'Exploring Uncertainty in Knowledge Representations', 109.

45 Gabel, 'Buckminster Fuller's World Game', 31.

46 Mihaly Csikszentmihalyi and Stith Bennett, 'An Exploratory Model of Play', *American Anthropologist* 73, no. 1 (1971), 46.

47 Gene Youngblood, 'Earth Nova', *Los Angeles Free Press*, 3 April 1974, 34. Quoted in Felicity D. Scott, 'Fluid Geographies: Politics and the Revolution by Design', in *New Views on R. Buckminster Fuller*, eds. Hsiao-Yun Chu and Roberto G. Trujillo (Stanford, CA: Stanford University Press, 2009), 160.

48 Bruno Latour, 'Why Gaia Is Not a God of Totality', *Theory, Culture & Society* 34, nos. 2–3 (2017), 75.

49 See Bruce Clarke, 'Rethinking Gaia: Stengers, Latour, Margulis', *Theory, Culture & Society* 34, no. 4 (2017), 3–26.

50 James Lovelock, *Gaia: The Practical Science of Planetary Medicine* (Oxford: Oxford University Press, 1991), 60.

51 Toby Tyrrell, *On Gaia: A Critical Investigation of the Relationship between Life and Earth* (Princeton: Princeton University Press, 2013), 4. Quoted in Latour, 'Why Gaia Is Not a God of Totality', 65.

52 Schlossberg, World Game Diary, 16.

53 W. Ross Ashby, *An Introduction to Cybernetics* (London: Chapman & Hall, 1958), 86.

54 René Thom, 'At the Boundary of Man's Power: Play', *SubStance* 8, no. 4 (1979), 12.

55 John McHale, 'Buckminster Fuller', in *The Expendable Reader*, ed. Alex Kitnick (New York: GSAPP Books, 2011),107.

56 Anonymous, 'World Game Pattern Recognition System for Resource Utilisation Planning', undated, Stanford University Collections M1090, Series 18: Project Files, World Game Subseries 2, box 27, Folder 2: World Game History, General (2 of 2), 2–3.

57 Mark Victor Hansen, 'An Aerial View of Ecology by World Game', Student Handbook for International Design Conference, Aspen 1970, Stanford University Collections M1090, Series 18: Project Files, World Game Subseries 2, box 14, 32/3: World Game History II, World Game Report (2 of 2).

58 Herbert Matter, *The World Game: Playing World Game (Getting power to the people)*, filmed at 1969 NY Studio School, World Game Seminar, Saturn Pictures, 23–30 June 1969. 60 min., b & w, 16 mm. Stanford University Collections M1090, Series 17: Subseries 7, 73i.

59 Schlossberg, World Game Diary, 20.

60 Anon. 'Handwritten notes on World Game Package', Stanford University Collections M1090, Series 18: Project Files, World Game Subseries 2, box 39, Folder 7, unpaginated.

61 Thomas Turner, Week-by-week schedule for World Game Workshop at SIU, Summer 1970 Stanford University Collections M1090, Series 18: Project Files, World Game Subseries 2, box 105, Folder 8: Tom Turner Files, Workshop.

62 Michael Ben-Eli and Edwin Schlossberg, Brochure for *The World Game for Government Executives*, 14–18 December 1970 and 1–5 March 1971, unpaginated. Stanford University Collections M1090, Series 18: Project Files, World Game Subseries 2, box 65, Folder 4: World Game Correspondence, Locations – Washington DC.

63 Brochure for *The World Game for Government Executives*, unpaginated.

64 List of Participants, The World Game Seminar, USDA Graduate School, 1–5 March 1971, Stanford University Collections M1090, Series 18: Project Files, World Game Subseries 2, box 65, folder 4.

65 Val Winsey, letter to Buckminster Fuller, dated 7 May 1972, Stanford University Collections M1090, Series 18: Project Files, World Game Subseries 2, box 39, Folder 18: World Game History, Val Winsey (3 of 3).

66 Val Winsey, 'World Game Has Serious Ends', *Pace Press*, Wednesday 11 March 1970, 10.

67 Winsey, 'World Game Has Serious Ends', 10.

68 Spencer Marks, 'Co-Op City Community Study', 1970, Stanford University Collections M1090, Series 18: Project Files, World Game Subseries 2, box 39, Folder 16: World Game History, Val Winsey (1 of 3).

69 Letter from Phillip H. Lawyer and Richard Meyer to Dr Marty Grober, 19 August 1970, Stanford University Collections M1090, Series 18: Project Files, World Game Subseries 2, box 65, Folder 4: World Game Correspondence, Locations – SIU, Free School.

70 Eric Berne, *Games People Play: The Psychology of Human Relationships* (New York: Grove Press, 1964).

3   Ecogame

In 1970, Thomas Turner, Director of Research and Development for the World Game at Southern Illinois University (SIU), sent architects Michael Uri Ben-Eli, Eytan Kaufman, and Michael Cutler to London to promote the World Game. At the first International Institute of Design Summer Session 70, organised by Alvin Boyarsky and hosted by the Bartlett School of Architecture from 13 July to 21 August 1970, the three proposed World Game workshops, but got no further than showing a video of Fuller. Sunwoo notes that the IID Summer Sessions, which ran from 1970 to 1973, 'provided refuge for theories and experiments that were without a professional or educational environment in which to develop'. Boyarsky's ambition was to establish an international network of avant-garde architects.[1] The following month, Ben-Eli presented at the Society for Academic Gaming and Simulation in Education and Training (SAGSET) first annual conference in Loughborough. Ben-Eli had joined the World Design Science Decade in 1963 whilst still a student at the Architectural Association in London and arranged the World Trends exhibits in Tuileries, Paris, in 1965, and in Bloomsbury Square, London, in 1967. Ben-Eli then began a PhD at the Institute of Cybernetics at Brunel University, where he studied under cybernetician Gordon Pask. In late 1969, Ben-Eli joined the World Game staff at SIU. A letter from Turner to the US Department of Justice to request a visa for Ben-Eli, dated 10 December 1969, identifies his 'exceptional merit as an architect, designer and general systems engineer'.[2]

At the SAGSET conference, Ben-Eli presented on the World Game. Cybernetician and systems engineer George Mallen, who was Chairman of SAGSET and co-founder of the Computer Arts Society (CAS), wrote to Turner on 15 September 1970 to thank him for sending Ben-Eli to the conference.

The society is very young and is very much concerned at the moment with defining possible strategies for using game ideas at all levels of education. Exposure to the World Game concept at this stage in our development therefore is particularly timely and will have far reaching effects.[3]

Ben-Eli and others from SIU also presented the World Game at the Computer 70 trade exhibition held in the Olympia, London, from 5 to 9 October that year. A letter from Kaufman to Turner, dated 25 July 1970, describes its enthusiastic reception.

Michael [Ben-Eli] arrived here last week and we had a world game presentation to an audience of about 45 people. It went well and there was some discussion about it later. A day after the presentation we met with a group of about 15 students, who were more interested in a workshop. We reviewed some of the slides again and answered more specific questions. One of the important questions that comes from European students is how to start a World Game unit in their own countries, like Finland for example. They are all looking for some 'instruction' manual from Carbondale.[4]

Figure 3.1 John Lansdown, diagram of Ecogame, Computer 70, Olympia, London, 5 to 9 October 1970, John Lansdown Archive, Middlesex University. Courtesy of Rob Lansdown.

At Computer 70, Ben-Eli, Kaufman, and Cutler likely would have encountered Ecogame, a dynamical systems simulation of a national industrial economy for multiple players, created by the CAS and shown in a 35-foot diameter dome (Figure 3.1). This chapter will discuss the background and ambitions of Ecogame and its three versions, first at Computer 70, then, in a different format, at the first symposium of the European Management Forum (precursor to the World Economic Forum) founded by economist Professor Klaus Schwab at Davos, Switzerland, from 24 January to 7 February 1971. After being bought, then rejected by International Business Machines (IBM) for its management centre in Blaricum, Belgium, a final version of Ecogame was installed at the Science Museum, London, in 1975, as part of the computer art booth in the Computing Then and Now gallery.

Mallen describes Ecogame as,

> The world's first interactive, multimedia, computer-controlled game and the result of an extraordinary collaborative team effort.[5]

As such, it was difficult to classify:

> on the one hand a management game, resource allocation game, teaching simulation, computer-controlled decision-making environment and on the other hand as an interactive multimedia computer-controlled game, procedural art, 'theatrical' performance, installation art etc.[6]

Before discussing Ecogame in more detail, I will provide some background to the CAS and to Mallen's investigations into computer-based modelling.

Computer Arts Society

Mallen, along with composer and engineer Alan Sutcliffe, architect John Lansdown, and artist Gustav Metzger founded CAS in late 1968 as a subgroup of the British Computer Society (BCS), following an informal session on computing in music at the International Federation for Information Processing Congress in Edinburgh in early August 1968. At the latter, Professor Stanley Gill suggested to Sutcliffe that he establish a group to study computer music, as Sutcliffe and Mallen worked together at the Electronic Music Studios in Putney. Sutcliffe expanded the group to include all the arts with the aim to 'spread more enlightened and intelligent use of computers, first in the arts and then more widely in society'.[7] The CAS held monthly meetings first at the Institute of Contemporary Arts (ICA), London, then at the BCS.

The BCS funded CAS 'to support pioneers by acting as an international forum for the exchange of ideas between people and by bringing them together for conferences, exhibitions, and monthly meetings'.[8] It aimed to bring artists, scientists, and engineers together to consider the (social) consequences of computing. By 1970, there were twenty-five members.

CAS members found an audience first through participation in the exhibition *Cybernetic Serendipity* curated by Jasia Reichardt at the ICA from 2 August to 20 October 1968, then with their own inaugural exhibition *Event One* in the Gulbenkian Hall of the Royal College of Art from 29 to 30 March 1969.[9] Mallen (with Pask), Sutcliffe, and Lansdown all participated in *Cybernetic Serendipity*, either as exhibitors or consultants.[10] *Cybernetic Serendipity* presented the work of over 130 participants, including composers, engineers, artists, mathematicians, and poets, to combine the science of control and communication with 'the faculty of making happy chance discoveries'. For Reichardt, the exhibition sought 'connexions between creativity and technology (and cybernetics in particular)'.[11] Judged by media reaction at the time, the principal achievement of *Cybernetic Serendipity* was to show that visitors could interact with computers and machines through play. As Leslie Stack of the ICA proclaimed, 'We want people to lose their fear of computers by playing with them and asking them simple questions'.[12]

A famous example of this interaction was Pask's *Colloquy of Mobiles*, a complex array of five robots, three 'female', two 'male', rotating on axes. In Pask's view, such 'aesthetically potent social environments', which were multi-modal communication systems, showed how to humanise technology through pleasurable dialogues and play or 'by interacting with the system at a higher level of discourse', similar to what Gregory Bateson called 'metacommunication' through play.[13] Such is the *Colloquy*'s importance, Maria Fernandez argues, that it 'May be discussed as a contribution to art, cybernetics, engineering, sociology, and artificial life'.[14]

The success of *Cybernetic Serendipity* helped the CAS to extend its membership. It also encouraged them to organise *Event One* with the support of Patrick Purcell, senior research fellow in the

Design Research Unit at the RCA. Mallen later joined the Design Research Unit as a part-time research fellow.

According to Catherine Mason, *Event One*

> Heralded the collaborative cross-disciplinary nature of working that came to signify the early period of media arts in Britain, where interactivity and process were as equally valued as object.[15]

In fact, CAS continued the cross-disciplinary collaboration promoted by the previous generation of cyberneticians, such as Pask, that computer or media artists followed. From 1949 to 1958, The Ratio Club in London, founded by John Bates of the Neurological Research Institute of the National Hospital, brought together psychiatrists, engineers, mathematicians, psychologists, and neurophysiologists to discuss cybernetics. Cyberneticians were often amateurs or disciplinary outliers, even as, on the one hand, cybernetics itself aimed to be integrative and interdisciplinary and, on the other hand, became an analytical toolbox widespread among military and industrial institutions.[16] Collaboration was then at the heart of Ecogame, with sculptors and programmers, behavioural scientists and architects, and many others coming together to exchange ideas in CAS meetings.[17]

Experiments in Modelling and Learning

Mallen had worked previously on modelling decision processes and was assistant to Pask, whom he joined at System Research Ltd in October 1964. Pask had founded System Research with Elizabeth Pask and Robin McKinnon-Wood in 1953.[18] In 1956, they had prototyped the first adaptive keyboard trainer for typists, which was then patented in 1961 as the Solartron Adaptive Keyboard Instructor (SAKI). SAKI, as it became known, adapted the difficulty and nature of learning tasks to the learner's performance. The more complex training machine, Eucrates, was devised by Pask in response to a request by Solartron Electronic Group for a 'nontrivial brain' to exhibit at the Physical Society Exhibition in London in 1956. Eucrates was for radar training but featured an assemblage of two machines, 'teacher' and 'pupil', in communication, which provided 'fundamentally an analogue simulation of a self-organising system'.[19]

Mallen describes System Research as 'a convergence point of cognitive science, computer technology and art'.[20] When he joined in 1964, Pask was working on a proposal for a 'cybernetic theatre'. After a meeting with Joan Littlewood, the playwright and founder of Theatre Workshop, Pask proposed a theatre wherein feedback loops between actors and audience would allow real-time plot development. Audience members could identify with a character and advocate for that character's course of action.[21] Pickering describes the proposal for a cybernetic theatre as 'another manifestation of Pask's ontology of open-ended performative engagement', an 'aesthetically potent environment' much like the later *Colloquy of Mobiles* but potentially more open-ended and dynamic.[22]

It was whilst working with Pask between 1966 and 1968 that Mallen became interested in 'the potential of computer simulation as an educational tool'.[23] In an article from 2013, Mallen discusses how he developed computer models of cognitive systems.

> I set about developing a model that would illustrate the dynamics inherent in the learning processes observed in adaptive teaching experiments at System Research Ltd.[24]

Mallen notes that despite Pask's initial wariness of digital modelling's simplifications, he later integrated such modelling into his theory of conversation.[25] Mallen developed the Learning Model (LMOD) as a dynamic hypothetical cognitive system that tested the operator to identify patterns between a display of coloured lamps and keys on the keyboard. The LMOD was a precursor to the Typist models of touch-typing and of the interacting learning systems of conversation theory. Mallen notes that such models were constructivist insofar as 'reality is constructed through a continuous process of neuronal/cognitive action' and were consistent with the insights of second-order cybernetics or neocybernetics.[26]

Ecogame was the product of this interest.

> I proposed that we make a simulation model of an economic system and use that as the basis for an interactive game in which players would make decisions and have the results fed back visually via the slide projectors.[27]

Two books from 1961 provided other key influences, Frank George's *The Brain as a Computer* and *Industrial Dynamics* by Jay Forrester.

Professor at the Sloan School of Management at MIT, Forrester, after working on early digital computers, had developed simulation models for industrial, corporate, and then urban dynamics since the late 1950s.[28] Forrester led Mallen away from the modelling of cognitive systems to modelling economies. Forrester's systems dynamics reached a wide audience when, famously, in the summer of 1970, the Club of Rome, a group of 'scientists, educators, economists, humanists, industrialists, and national and international civil servants' led by Italian industrialist Aurelio Peccei, employed Forrester's modelling techniques to address the 'world *problématique*', or global problems such as continued growth in a closed system of finite resources.[29] The report of the Club of Rome, published in 1972 as *The Limits to Growth: A Report on the Club of Rome's Project on the Predicament of Mankind*, became widely popular and frequently criticised for its projection of civilisational collapse by the year 2050 given current demographic and economic trends. Derided as simplistic, biased, and data poor by most scientists, economists, and politicians, world dynamics modelling nonetheless has enjoyed a continued following in the environmental movement and among advocates of sustainable development and, more recently, degrowth. As Paul Edwards remarks, 'although world dynamics failed as a scientific enterprise, it succeeded as an intervention in political culture', introducing global simulation modelling and managed sustainability into the public sphere.[30] *Limits to Growth* and the subsequent work of the Club of Rome also advocated for supranational governance of these planetary systems, based upon long-term, holistic thinking, dynamic trend analysis, and projective, counterfactual modelling.[31]

Forrester was criticised for favouring the structure and dynamics of a model over the accuracy and comprehensiveness of a data set. Indeed, Forrester, and the Systems Dynamics Group that produced world models for *Limits to Growth*, led by Dennis and Donella Meadows (the former having been Forrester's student), often guessed data used to parameterise their models. To be fair, comprehensive, consistent, and disaggregated data sets were extremely hard to come by at the time (a problem faced also by meteorologists and climatologists, as Paul Edwards notes), or simply did not exist. It is understandable, perhaps, that with only messy data sets available, the model's structure and dynamics took precedence. What is more, Forrester produced models for managers and politicians rather than scientists, and the former often had to make decisions based on incomplete data.

The systems modelled by Forrester were complex, 'multiloop nonlinear feedback systems'.[32] This was especially true of social systems and of the interaction of these latter with technical systems and the natural environment in a 'world system', which meant that they could not be modelled in isolation from other systems and interventions in those systems could have unpredictable and deleterious consequences, especially through positive feedback loops.[33] Humans struggle to interpret and understand the counterintuitive qualities of such systems, Forrester claimed, but computer modelling would overcome these human limitations to 'determine the dynamic consequences when the assumptions within the model interact with one another'.[34]

Forrester's *Urban Dynamics* of 1969 proposed a dynamic model of a city or 'urban system' comprised of the three subsystems of industry, housing, and people. Cybernetician and Club of Rome member Alexander Christakis reviewed *Urban Dynamics* in 1970, noting that the model was instructive of functional relationships and policy consequences despite not being 'a good representation of the real world'.[35] Further criticisms were that the system boundary of Forrester's urban model was unclear as was its correlation to its environment. In 1971, Forrester scaled up this modelling in *World Dynamics*.

Retrospectively, Forrester cautioned that modelling an organisation and advising its managers is rarely enough to change behaviour. Not least, managers did not always accept the model's version of consequences. 'Exposure to dynamic thinking' should begin at an early age (from ten years old) and include learners in modelling, Forrester urges.[36]

Ecogame

Ecogame was an early attempt to expose the public to 'dynamic thinking' through its model upscaled from urban dynamics to an 'artist's impression' of a national economy. According to Mallen, Ecogame modelled

> a very precisely defined situation presented as an extremely simplified and naïve picture of the way in which wealth is distributed through our social and industrial systems ... an artist's impression of an economy – simple and engaging.[37]

Although its model was not as complex as those developed by Forrester, players still might learn about the consequences of decisions made collectively or unilaterally concerning multiple variables. In the Game's first iteration at Computer 70, these consequences were displayed to players as selections of 35 mm slides (each projector held a carousel of 80 slides) projected onto the interior of a 35-foot diameter dome. These slides offered players real-time feedback of 'the state of various parameters in the game', especially the resources available, and reflected the status or 'mood' of the national economy in play according to whether these resources were depleted or abundant. 'The overall resource level in the game', Mallen explained, 'is taken as a measure of the state of our natural environment'.[38] If the economy were performing well, there would be slides of abundance and happy citizens. If not, slides would show dole queues, civic unrest, and environmental degradation.

> The game monitored the performance of all players and prevented too much profit flowing to any one player by equalising distribution parameters whenever the measured inequality exceeded a specific threshold.[39]

This meant that capital accumulation would be halted beyond a certain threshold by redistribution of wealth through revolution. Implicitly, at least, Ecogame took a Marxist approach to political economy.

The prefix 'eco', then, maintained its etymological root of *oikos*, the proper management of a household and its resources. However, like the World Game, this household was scaled up, to the nation state if not to the planet. Ecogame showed starkly how mismanagement through private accumulation led to ecological and political collapse.

Upon entry, players received an instruction booklet detailing options and gameplay. Teams of three players decided how to draw from a 'main resource reservoir' and to distribute wealth within the system, according to the variables of personal motivation, common wealth, investment returns, differential distribution of income, and resource availability. Resources could be consumed or, through investment, returned to the reservoir (Figure 3.2). Each player was responsible for either urban, social, or industrial development and had to manage flows of resources and information. Each decision point was presented as a four-fold choice or four responses to a problem, which often opposed maximum profit to social cost. The interaction of players' decisions produced a complex, unpredictable system in play, where the relative success of private accumulation or collective wealth was 'As in real life ... difficult to define'.[40] Lansdown notes that Ecogame produced 'complex and nonlinear' interaction and

48   *Ecogame*

252   *Programmed Learning and Educational Technology, Vol. 10, No. 4, July, 1973*

Figure 1

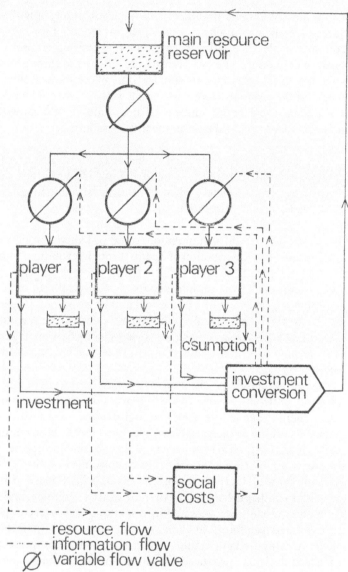

Figure 3.2 George Mallen, Ecogame Flow Model, 1970. Courtesy of Sarah Mallen.

'takes into account not only the decisions made by the players as individuals but also those made by the participants as a whole'.[41] At the time, Mallen noted that the game structure of Ecogame could be modified and interpreted in several ways, such as to allow players to 'select the context within which they make decisions'.[42] In a statement written a week prior to Ecogame's exhibition at Computer 70, Foundation General Systems Ltd (Anthony McCall, Christine McNulty, and John McNulty), who were contracted by CAS to implement the design, explained its three ambitions: first, to show how an integrated system might display emergent behaviour; second, to present a future, decentralised work scenario facilitated by telecommunications and computational power; and third, to establish an experiment in systems management and resource allocation. Their principal concerns were systems integration and the development of a multi-modal, portable communications device.[43]

The Game Environment

With Ecogame, as with the World Game proposals discussed in Chapter 1, a dome provided an immersive architecture for intense experience and instruction. The dome was preferred also because it provided an interface of planet as lifeworld (sphere) and as object (globe). A plan and elevation of the proposed dome at Computer 70 shows a 'game configuration of three teams and four players'.[44] From a central plinth, three arrays of projectors project at 120 degrees from each other onto the upper interior wall. Below each projection, around the perimeter of the dome, three arc desks each seat four players, who look up at the consequences of their team's decisions. In the middle of the floor, visitors gather to observe the game. An irregular octagonal wall encloses the dome, producing an entrance area with two tables, where, presumably, players and visitors could find instructional material for the game.

Filmmaker Anthony McCall, who selected the slides for projection, described the cool, blue atmosphere inside the dome and the rhythmic clicking of the slide carousels. Each solution to a problem carried a social score and an individual score.

The game was run on a PDP 10 computer on Great Portland Street, with an Idiom Graphic Display 'acting as Umpire', that is, assessing the 'balance of personal gain and social benefit' and selecting the appropriate slide sequence to represent this balance.[45] Chris Winter, who had joined Compugraphics International (Aldershot) in September 1969, and worked on Ecogame, gave a more detailed technical description, noting that the video terminals were connected to a PDP mainframe in Switzerland, which then fed back the results of players' decisions to be displayed on a graphics terminal run by a minicomputer with 8k of memory. 'When I look back at that, that it was quite stunning for 1970; it was quite leading edge'.[46]

European Management Forum

Ecogame showed how computer art might experiment with learning technologies for economic and environmental management. This brought it to the European Management Forum (EMF) symposium in Davos, Switzerland, where it ran with the same software and game structure but without the geodesic dome (Figure 3.3). Instead, it took place in an auditorium of the Kongresshaus in Davos as part of the conference's educational support programme. The Forum's broad ambition was to elaborate 'role models and concepts for responsible and successful management', based upon Schwab's 'stakeholder theory' of management.[47] The symposium used the latest information and communications technology, with Ecogame as its central learning technology.

> Computer generated models were employed to analyse the implications of strategies under consideration and predict the impact that any specific allocation of resources would have on businesses and the environment. Monitors displayed the managers' 'decisions', while colour slides illustrating the consequences of those choices were projected on large screens.[48]

Figure 3.3 George Mallen, Ecogame installation at the European Management Forum, Davos, Switzerland, 24 January to 7 February 1971. Courtesy of Sarah Mallen.

Due to its success, Ecogame might have influenced cybernetician Stafford Beer (who attended the EMF symposium) to develop its real-time feedback technology for the Operations Room of Cybersyn, a project in cybernetic governance developed from 1971 to 1973 under socialist Chilean president Salvador Allende. Mallen recalls,

> Stafford Beer was there and as a result the slide projection technology was used as part of the support infrastructure for the famous decision room for industrial planning in Allende's government in Chile.[49]

However, this influence upon Beer's 'liberty machine', which applied his Viable Systems Model to Chile's newly nationalised core industries in an effort to match technocratic management with worker participation, has left little trace in the literature. Gui Bonsiepe led the Industrial Design Group at the government research centre, the State Technology Institute or INTEC, in Santiago, in designing the octagonal Operations Room. Beer requested something akin to a gentlemen's club (hence the inclusion of a bar), but otherwise, as one of the designers, Rodrigo Walker, asserted, 'There was no reference point for this project'.[50]

Following the EMF symposium, IBM bought Ecogame from Mallen's System Simulation Ltd. to use at its management centre in Blaricum, The Netherlands. As noted previously, IBM had long supported post-war experiments in immersive 'information machines' (dome architectures upon whose interior surfaces dynamic slide compositions were projected) for public education, or multi-media education centres for 'worldliness and citizenship'.[51] In Ecogame, however, IBM invested in a learning tool for its own staff rather than a more spectacular architectural advertisement for its products. Nevertheless, IBM soon rejected Ecogame because the former's senior management refused to accept that their economic model could have damaging political and environmental consequences. 'IBM said senior management felt the game attacked their value system', George Mallen declared. The problem was likely the political economy implicit in Mallen's 'income maldistribution check'. IBM middle management enjoyed Ecogame, Mallen claimed, but senior management

> could not cope with the fact that seemingly reasonable business decisions had unacceptable political and environmental consequences. Rather than admit their preconceptions might be wrong, the game was scrapped.[52]

The result of this rejection was that Ecogame was installed in 1975 at the Science Museum in London, where it returned to the domain of computer art. Its period of what Frieder Nake called 'technocratic dadaism' seemed to have come to an end.[53]

Computing Then and Now

The computer art booth in the Computing Then and Now gallery at the Science Museum was a darkened tunnel on an east-west axis between pillars at the east end of the gallery, positioned between the analogue and digital sections. In the darkness were up to eight exhibits, including two videos, seven audio compositions on tape, and four slide shows, one of which, *Machines that Make Art*, introduced the booth's contents and summarised artists' ambitions to test the possibilities and limitations of computers. Swade concludes,

> The contents of the Science Museum's computer art booth serve as an index to attitudes, aspirations, hardware, practice, and the creative products of the computer art movement in the mid-1970s. In straddling two cultures [science and art], literally and metaphorically, the computer art booth was an anomaly. An intriguing one.[54]

Although a plan of the Computing Then and Now exhibition is available, there are no details of how the Ecogame was presented in the booth.[55]

Given its evolution from Computer 70 to EMF to IBM, what does it mean to present Ecogame as computer art at the Science Museum? It exceeded several of the roles of computer art as artist and critic Brian Reffin Smith understood them in 1975, except for prioritising procedure over design.[56] Lansdown includes Ecogame as an example of 'procedural' art. Far from equating computer art with computer graphics, where computation provides an artistic tool, for Ecogame 'the object is the process'.[57] Frieder Nake, writing of the 'algorithmic principle' of computer art, notes that when the computer becomes the medium rather than a tool, 'interactivity enters its algorithmically based form'.[58] Ecogame offered a training in algorithmic interactivity in the service of management for collective benefit.

On the other hand, compared to the World Game Seminar, which, in the absence of 'optimum technological facilities', required what Peter Weibel has called the 'intuitive' rather than 'exact' application of algorithms, Ecogame's gameworld required a more exact algorithmic interactivity.[59] It took a step further than the World Game toward these facilities, but this meant that its gameplay was more exactly rule-based. In both games, of course, the logic of selection assumes a player's acceptance of, in Bateson's analysis, certain rules and conceptual premises, which therefore make it difficult for the players to conceive that there might be other ways of meeting and dealing with each other.[60] Ecogame's automation of this logic, compared to the more negotiated and open logic of the less computationally sophisticated World Game Seminar, might suggest a less favourable evaluation of the former. Ecogame did not integrate live data and its modelling did not occur in real time. Instead, decisions and their consequences played out in a gameworld that was already weighted by its designers, the parameters of its simulation largely unresponsive to its players' interactions, experiences, and knowledge. Nevertheless, the biases designed into Ecogame, notably the parameter restricting private accumulation, showed that, for its designers, modelling, simulation, and designed interaction were political from the start. This contrasts with the World Game designers' and players' belief that computational power would render politics obsolete, expressed succinctly in the title of one of Herbert Matter's films of the Seminar: 'Politicians will yield to the computer'.[61]

Notes

1 Irene Sunwoo, 'Pedagogy's Progress: Alvin Boyarsky's International Institute of Design', *Grey Room* 34 (Winter 2009), 30.
2 30, Folder 5: World Game History 2, Personnel 1969*
3 62, Folder 2: World Game Correspondence, Locations, London
4 62, Folder 2: World Game Correspondence, Locations, London
5 George Mallen, 'Ecogame: Computing in a Cultural Context', unpublished draft paper, September 2019, unpaginated.
6 Mallen, 'Ecogame: Computing in a Cultural Context'.
7 Alan Sutcliffe, 'Patterns in Context', in *White Heat Cold Logic: British Computer Art 1960–1980*, eds. Paul Brown, Charlie Gere, Nicholas Lambert, and Catherine Mason (Cambridge, MA: MIT Press, 2009), 178.
8 Catherine Mason, 'The Fortieth Anniversary of Event One at the Royal College of Art', paper delivered at Electronic Visualisation and the Arts annual conference, 6 to 8 July 2009, Published

online by the BCS via ScienceOpen: https://www.scienceopen.com/hosted-document?doi=10.14236/ewic/EVA2009.15, doi:10.14236/ewic/EVA2009.15.

9 Brian Reffin Smith, 'Computer Art: Recent Trends', *Computer Aided Design* 7, no. 4 (October 1975), 225.

10 Mason, 'The Fortieth Anniversary of Event One'.

11 Jasia Reichardt, 'Cybernetics, Art and Ideas', in *Cybernetics, Art and Ideas*, ed. J. Reichardt (Greenwich, CT: Graphic Society, 1971), 11.

12 Leslie Stack quoted in Linda Talbot, 'Meet the Friendly Robots', *Hampstead and Highgate Express*, 26 July 1968. For a critical assessment of the media reception of *Cybernetic Serendipity*, see Rainer Usselmann, 'The Dilemma of Media Art: Cybernetic Serendipity at the ICA London', *Leonardo* 36, no. 5 (2003), 389–396.

13 Gregory Bateson, 'The Message —This Is Play,' *Transactions of the Second Conference on Group Processes*, October (1955), 145–242, and 'A Theory of Play and Fantasy,' *Steps to an Ecology of Mind: Collected Essays in Anthropology, Psychiatry, Evolution, and Epistemology* (Chicago, IL: University of Chicago Press, 1972).

14 Maria Fernandez, '"Aesthetically Potent Environments", or How Gordon Pask Detourned Instrumental Cybernetics', *White Heat Cold Logic: British Computer Art 1960–1980*, eds. Paul Brown, Charlie Gere, Nicholas Lambert and Catherine Mason (Cambridge, MA: MIT Press, 2009).

15 Mason, 'The Fortieth Anniversary of Event One'.

16 See Owen Holland and Phil Husbands, 'The Origins of British Cybernetics: The Ratio Club', *Kybernetes* 40, nos. 1/2 (2011), 110–123. Also, Bronac Ferran and Elisabeth Fisher, 'The Experimental Generation: Networks of Interdisciplinary Praxis in Post War British Art (1950–1970)', *Interdisciplinary Science Reviews* 42, nos. 1–2 (2017), 1–3.

17 See John Lansdown, 'The Name of the Game Is...? A Personal View of the Computer Arts Society's Project', *The Computer Bulletin* 14, no. 9 (September 1970), unpaginated.

18 See Elizabeth Pask, 'Today Has Been Going on for a Very Long Time', *Systems Research* 10, no. 3 (1993), 143–147 and Robin McKinnon-Wood, 'Early Machinations', *Systems Research* 10, no. 3 (1993), 129–132.

19 F. F. Kopstein and Isabel J. Shillestad, A Survey of Auto-Instructional Devices, Aeronautical Services Division Technical Report 61-414, September 1961, AD 268223.

20 Mallen, 'Ecogame: Computing in a Cultural Context'.

21 Gordon Pask, 'Proposals for a Cybernetic Theatre', 1964, privately circulated monograph, Theatre Workshop and Systems Research, held at Pask Archive: http://www.pangaro.com/Pask-Archive/.

22 Pickering, *The Cybernetic Brain: Sketches of Another Future* (Chicago, IL: Chicago University Press, 2010), 350.

23 Mallen, 'Ecogame: Computing in a Cultural Context'.

24 George Mallen, 'Early Computer Models of Cognitive Systems and the Beginnings of Cognitive Systems Dynamics', *Constructivist Foundations* 9, no. 1 (2013), 137.

25 Gordon Pask, *Conversation Theory: Applications in Education and Epistemology* (Amsterdam: Elsevier, 1976).

26 Mallen, 'Early Computer Models of Cognitive Systems', 138. See Bruce Clarke and Mark B. N. Hansen, 'Introduction: Neocybernetic Emergence,' in *Emergence and Embodiment: New Essays on Second-Order Systems Theory*, eds. B. Clarke and Mark B. N. Hansen (Durham, NC: Duke University Press, 2009).

27 Mallen,'Ecogame: Computing in a Cultural Context'.

28 See Jay Forrester, *Industrial Dynamics* (Waltham, MA: Pegasus, 1961).

29 Donella Meadows, Dennis Meadows, Jørgen Randers, and William W. Behrens III, *The Limits to Growth: A Report on the Club of Rome's Project on the Predicament of Mankind* (New York: Universe Books, 1972), 9.

30 Paul N. Edwards, *A Vast Machine: Computer Models, Climate Data and the Politics of Global Warming* (Cambridge, MA: MIT Press, 2010), 370–371.

31 See Alexander King and Bertrand Schneider, *The First Global Revolution* (New York: Pantheon, 1991) and Dennis Meadows, Donella Meadows, and Jørgen Randers, *Beyond the Limits: Global Collapse or a Sustainable Future* (London: Earthscan, 1992).

32 Jay Forrester, 'Counterintuitive Behavior of Social Systems', *Simulation* 16, no. 2 (1971), 61.

33 Jay Forrester, *World Dynamics*, Second Edition (Cambridge, MA: Wright-Allen Press, 1973), 5.

34 Forrester, 'Counterintuitive Behavior of Social Systems', 62.

35 Alexander Christakis, review in *Technological Forecasting* 1 (1970), 427.

36 Jay Forrester, 'The Beginning of System Dynamics', Banquet Talk at the International Meeting of the System Dynamics Society, Stuttgart, 13 July 1989, 13: https://web.mit.edu/sysdyn/sd-intro/D-4165-1.pdf

37 Mallen, 'Ecogame: Computing in a Cultural Context'.

38 George Mallen/Computer Arts Society, 'Ecogame', September 1970, 3, Stanford University Collections M1090, Series 18: Project Files, World Game Subseries 2, box 62, Folder 2: World Game Correspondence, Locations, London.

39 Mallen, 'Ecogame: Computing in a Cultural Context'.

40 Mallen, 'Ecogame: Computing in a Cultural Context'.

41 John Lansdown, 'Computer Graphics ≠ Computer Art', *Page* 19 (December 1971), 2.

42 Mallen/Computer Arts Society, 'Ecogame', 2.

43 Foundation General Systems Ltd, 'A Description of the Computer 70 Theme Exhibit, Olympia, London 5–9 October 1970', *PAGE* 16, June 1971, unpaginated.

44 George Mallen/Computer Arts Society, 'Ecogame', 3.

45 Anthony McCall, PAGE16, June 1971.

46 Tom Abram, interview with Chris Winter, 9 January 2020 https://archivesit.org.uk/interviews/chris-winter/

47 World Economic Forum, *World Economic Forum—A Partner in Shaping History, The First 40 Years, 1971 to 2010* (Cologny: World Economic Forum, 2009), 6. http://www3.weforum.org/docs/WEF_First40Years_Book_2010.pdf.

48 World Economic Forum, *World Economic Forum*, 9.

49 Mallen, 'Bridging Computing in the Arts and Software Department', 197–198.

50 Walker quoted in Eden Medina, *Cybernetic Revolutionaries: Technology and Politics in Allende's Chile* (Cambridge, MA: MIT Press, 2011), 121.

51 Justus Nieland, 'Midcentury Futurisms: Expanded Cinema, Design, and the Modernist Sensorium', *Affirmations: Of the Modern* 2, no. 1 (2014), 56.

52 Patrick Ryan, 'Not Playing the Game', *New Scientist* 57, no. 828 (11 January 1973), 96.

53 Frieder Nake, 'Technocratic Dadaists' *PAGE* 21 (March 1972), unpaginated.

54 Doron D. Swade, 'Two Cultures: Computer Art and the Science Museum', in *White Heat Cold Logic: British Computer Art 1960–1980*, eds. Paul Brown, Charlie Gere, Nicholas Lambert, and Catherine Mason (Cambridge, MA: MIT Press, 2009), 216.

55 Swade, 'Two Cultures', 216. For the plan, see Tilly Blyth, 'Narratives in the History of Computing: Constructing the Information Age Gallery at the Science Museum', in *Making the History of Computing Relevant*, eds. A. Tatnall, T. Blyth, and R. Johnson. HC 2013. IFIP Advances in Information and Communication Technology 416 (2013). doi:10.1007/978-3-642-41650-7.

56 Reffin Smith, 'Computer Art: Recent Trends', 225.

57 Lansdown, 'Computer Graphics ≠ Computer Art', 2.

58 Frieder Nake, 'The Semiotic Engine: Notes on the History of Algorithmic Images in Europe', *Art Journal* 68, no. 1 (Spring 2009), 88.

59 See Peter Weibel, 'It Is Forbidden Not to Touch: Some Remarks on the (Forgotten Parts of the) History of Interactivity and Virtuality', in *Media Art Histories*, ed. O. Grau (Cambridge, MA: MIT Press, 2007), 23.

60 Gregory Bateson, letter to Norbert Wiener, cited in William Poundstone, *Prisoner's Dilemma* (New York: Anchor, 1993), 168. See Pamela Lee, *New Games: Postmodernism after Contemporary Art* (London: Routledge, 2013), 131.

61 Herbert Matter, *World Game Can Work (Politicians Will Yield to the Computer)*, filmed at 1969 NY Studio School, World Game Seminar, Saturn Pictures, 23–30 June 1969. 60 min., b & w, 16 mm. Stanford University Collections M1090, Series 17: Subseries 7, 73j.

4 Ecologies of the Future

By the time the newly married John McHale and Magda Cordell McHale moved from England to Carbondale, Illinois, in 1961 to work with Fuller on the World Resources Inventory of Human Trends and Needs (WRI, begun in 1959) and then the World Design Science Decade (WDSD) (1965–1975), John McHale already knew Fuller's work in detail. He had written an article on Fuller in 1956 in *The Architectural Review* and had invited Fuller to a lecture at the Institute of Contemporary Art, London, in June 1958. In 1962, John McHale would write the first monograph on Fuller, published by Prentice Hall. Nevertheless, the move to the United States transformed the McHales' work. Although these two founding members of the Independent Group continued to produce art privately, following their move, Magda recalled in 1978, 'for both of us, just to produce art was not enough'.[1] The McHales explored instead 'the future imperative', which marked

> A stage in human global development at which the continuous review and assessment of the long-range future implications of our past and present actions becomes crucially important for the survival of human society.[2]

They established several future studies centres, beginning with the Center for Integrative Studies at SUNY Binghamton (CIS Binghamton) in autumn 1969, after moving there the previous year, with John as Director and Magda as one of two research assistants. In 1975, they wrote of an emerging 'cultural collage', 'a mesh of horizontally interlinked networks' and institutions that would 'optimise diversity' from the bottom up.[3] Their own future studies centres promoted this cultural collaging, taking the World Game as a prototype for the *'flexibly dispersed* learning centres … interlinked to centralised libraries and other major facilities' required for design science and planetary ecology.[4] The McHales collaborated equally on all their projects, despite the ongoing prominence of John in the scant literature that discusses their futurological work. As interviewer Gay McFarland summarised in 1978, 'They say that people who like each other, who work well together, are more productive. And that neither of them would have the energy or the drive without the other'.[5]

John McHale was Director of the World Resources Inventory for the WDSD from 1963 to 1967, replaced in June 1968 by Carl G. Nelson, who had been Senior Research Administrative Assistant and Projects Coordinator at WRI.[6] One of the WRI's principal objectives was to 'act as a support data system for the World Game/Spaceship Earth research activity'.[7] Data collection, analysis of trends, and futures modelling then became the principal roles of the McHales' future studies centres.

A brief survey of their activities through the 1970s gives a sense of their status and influence. In 1970, John McHale served as the only non-government official on the Temporary State Commission on the Environmental Impact of Major Public Utility Facilities, appointed by New York Governor Nelson Rockefeller.[8] By December 1971, he was a Fellow of the World Academy of Art and Science, the New York Academy of Sciences (effective from 1 December 1971), the Royal Society of Arts, and the American Geographical Society.[9] They gained significant funding for futures research. In September 1971, the CIS Binghamton was awarded $70,000 from the National Science Foundation for the research project 'Projected Relationships Between Energy and Resource Requirements'. The study, which involved undergraduate and graduate students in the Center, examined technological solutions that matched population with resources.

> This analysis will lead to the projection of the severity of the problems as they relate to time. Priorities and alternative strategies will be evaluated within the context of overall resources requirements for different time periods.[10]

John McHale also served as consultant to the United Nations (UN) Environment Programme, the UN Conference on Science and Technology for Development, and the Aspen Institute Program in International Affairs.[11] In September 1973, he opened the five-day Rome Special World Conference on Futures Research, titled 'Human Needs-New Societies-Supportive Technologies', with a review of futures literature in which he lamented planning by 'one election minds' and asked how incomplete, sometimes misleading models (he names Forrester and Meadows) might 'offer a net gain in perception and insight'. In a world that is 'becoming less deterministic', he enthused, 'we are more than ever in charge of our own destinies'.[12] Also at the conference, Magda joined a panel on art as an indicator of the future. John then presented the report 'Changing Information Environment: Policy Consequences and Implications' at the International Computer and Telecommunication Conference in Milan.[13] The following year, the UN Institute for Training and Research granted the CIS Binghamton $20,000 for futuristic research, the first time the Institute had awarded a grant to a non-UN agency.[14] This research produced an international survey of future studies projects and centres, published first in *Ekistics* in 1976, then, in 1977, as *Futures Directory: an international listing and description of organizations and individuals active in futures studies and long-range planning*.[15] The World Society of Ekistics elected John McHale to its membership in September 1974.[16] In 1973, Magda was Senior Research Fellow at the East-West Center in Honolulu, a hub of futures research, and in 1977, she was Presidential Fellow at the Aspen Institute for Humanistic Studies in Colorado. For International Women's Year conference in Mexico in 1975, the McHales wrote (with Guy Streatfield as researcher) *Women in World Terms: Facts and Trends*.[17]

As noted, the McHales separated private art from public research, but in several respects, their futurology and design science expanded investigations into information, technological futures, mass culture, gender, and lifestyle that the two first expressed through collage and painting in the 1950s.[18] They used charts, graphs, and diagrams as 'ikons' to promote 'planetary housekeeping' and 'ecological redesign'.[19] John McHale's writings show this evolution of their work. In one of his better-known early essays, 'The Expendable Ikon', published in the February and March 1959 issues of *Architectural Design*, he noted that the fine arts had ceded their traditional role of making 'statements about man's total environmental situation' to the ikons of the 'technological folks arts' (TV, radio, advertising, news media). By 1967, he asserted that art had to define 'alternative cultural strategies, through a series of communicative gestures in multi-media forms'. Life was now to be defined as art, 'as the only contrastingly permanent and continuously unique experience'.[20] By the end of the 1970s, the McHales directed their activities almost entirely towards an 'exploration and definition of social and cultural directions'.[21] They answered the future imperative by detailing the 'externalised metabolic systems of humanity' (natural, technological, and infrastructural)[22] to allow 'a synoptic view of world trends and alternative futures'.[23] They inventoried vast amounts of data on past and present trends, appropriated military and technocratic modelling techniques, such as Cross Impact Analysis, Delphi Method, Gaming, Network Analysis, Probabilistic Forecasting, and Scenario Building[24] for egalitarian ends, such as to give 'explicit attention to the voice or place of women ... in the debate [about global problems], in the locus of the problems, [and] in the ways in which solutions are to be implemented'.[25] Despite this evolution of interest from 'technological folks arts' to futurology and world architecture, the later work of the McHales 'is still under-explored'.[26]

Expanded Collage

We can follow this evolution in the use of collage. Each of John's charts, graphs, and diagrams for the McHales' futurological publications is 'a polemical artwork, an image in its own right, built on verifiable state-of-the-art data but crafted to particular aesthetic effect', Mark Wigley writes.[27] Anthony Vidler describes a 'sociologist-collagist', whose intellectual career naturally shifted from 'collage art to systems analysis, from art to science'.[28] Consider an example of such collage art for informational purposes. In 1955, for the cybernetician E. W. Meyer's address to the IG at the Institute of Contemporary Arts (ICA), 'Probability and Information Theory and their Application to the Visual Arts', John McHale made a first foray into information design, an example of which is the collage *Forms of Coding* (Figure 4.1). He recalled that having found an expert on cybernetics, his ideas would have to be explained to the IG through several diagrams.

> We had a standard Shannon diagram, then we had an example of coding, all laid out, we had a statistical probability ... I think there were five sets of diagrams.[29]

The previous year, he had already begun to make collages based upon information processing, such as *Transistor* (1954), which reworked Claude Shannon's famous diagram from his 'A Mathematical Theory of Communication', published in 1948. Jacquelynn Baas describes *Transistor* as a 'visual equivalent for the processing

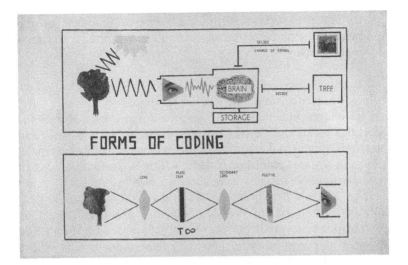

Figure 4.1 John McHale, *Forms of Coding*, 1955, from John McHale Archive, black and red fibre-tip pen. YCBA 2798443-0001. Yale Center for British Art, Gift of Magda Cordell McHale. © Yale Center for British Art.

of information',[30] which showed John McHale's interest in how data is processed, ordered, and reconfigured through electronic and organic systems. Although *Transistor* lacks the visual unity and simplicity of Shannon's diagram and exploits the fragmented character of collage, it would be a mistake to assume that this introduces noise into the channel, so to speak, of scientific communication.[31] Instead, he seems to show how alike are collage and diagram. A diagram provides a 'disunified field of presentation' that can be apprehended sequentially and from multiple vantage points at different scales.[32] This is in large part what breaks the diagram's iconic relation to an object and allows it to model abstract entities such as processes, relations, directions, and functions. This disunity allows for the layering and separation, or parallelism, advocated by Edward Tufte, which stratifies data and allows for communication of the proper relation among variables.[33] Finally, collage is a dynamic medium. In his statement for an exhibition of collages at the ICA in 1958, *3 Collagists – new work by E. L. T. Mesens, John McHale, Gwyther Irwin*, John McHale notes that collage is the most appropriate medium for ikons derived from mass media.

[Collage is] parallel to our own experience of the image, as we turn the pages of a magazine, watch a movie, or scan a newspaper, fragments perceived at random are organised into meaningful wholes.[34]

In his collages for the Meyer talk, John McHale shows clearly how collage can be diagrammatic and informational.

This interest in collage as information design continues into the McHale's futurology. By the early 1970s, John McHale aspired to match the graphs and diagrams of such as the Club of Rome report, *Limits to Growth*, published in 1972, which he praised as 'a set of poetic metaphors instantly available on a world scale'.[35]

John McHale first published several of his own 'poetic metaphors' in the series of WDSD documents produced from 1963 to extrapolate trends from the World Resources Inventory and to promote design science. The first of these documents, Phase I (1963), Document 1: Inventory of World Resources, Human Trends and Needs, features, for example, a squared spiral showing the geological timescale of 'Earth/Man' from the birth of the Earth to the Quaternary Period. Other diagrams show 'Man's Increasing Vertical Mobility', the industrial lifecycle of metals, and world maps showing the distribution of 'energy slaves per capita' and 'calories per capita per day per area'. These diagrams visualised data collected and interpreted from a vast literature that ranged across the sciences and humanities. They were generative, too, in correlating variables, showing how, for example, the relative size of the planet had shrunk due to travel and telecommunications and correlating this shrinkage to a series of technological revolutions. This visual correlation provided evidence that planetary technologies would meet planetary ecological challenges.

At the point, then, where man's affairs reach the scale of potential disruption of the global ecosystem, he invents, with seeming spontaneity, precisely those conceptual and physical technologies which enable him to deal with the magnitude of complex planetary society.[36]

Diagrams were such a 'conceptual technology', as they integrated environmental and human systems, the latter separated into biophysical, psychosocial, and technological. Beneath the 'Global Ecosystem' diagram (Figure 4.2), a flow diagram to

THE GLOBAL ECOSYSTEM

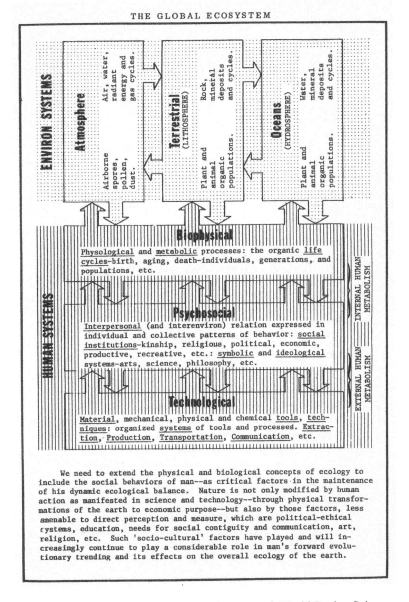

We need to extend the physical and biological concepts of ecology to include the social behaviors of man--as critical factors in the maintenance of his dynamic ecological balance. Nature is not only modified by human action as manifested in science and technology--through physical transformations of the earth to economic purpose--but also by those factors, less amenable to direct perception and measure, which are political-ethical systems, education, needs for social contiguity and communication, art, religion, etc. Such 'socio-cultural' factors have played and will increasingly continue to play a considerable role in man's forward evolutionary trending and its effects on the overall ecology of the earth.

Figure 4.2 John McHale, 'The Global Ecosystem', World Design Science Decade 1965–1975, Phase II (1967), Document 6, The Ecological Context: Energy and Materials, World Resources Inventory, ILL. 1967, 41. Courtesy of John McHale and Magda McHale Archives Foundation.

correlate the constituent subsystems of a planetary system, John McHale wrote,

> We need to extend the physical and biological concepts of ecology to include the social behaviours of man – as critical factors in the maintenance of his dynamic ecological balance. ... 'Socio-cultural' factors have played and will increasingly continue to play a comfortable role in man's forward evolutionary trending and its effects on the overall ecology of the earth.[37]

John McHale republished this diagram several times. It appeared in his article 'Global Ecology: Toward the Planetary Society', published in 1968, then in his two monographs, *The Future of the Future* (1969) and *The Ecological Context* (1970), where it supported the broader ambition of establishing the 'physical operational parameters for the planet—the ecological or housekeeping rules that govern human occupancy' and prioritising a planetary, rather than local, approach to 'ecological health'.[38] 'There are no "local" problems anymore', he argued, 'such as may be left to the exigencies and dangerous predilections of local economic or political "convenience".[39]

Diagrams were an integrative, dynamic, and multi-scalar informational medium that provided images of a collaged world architecture whose raw material was not building so much as statistics. Indeed, the McHales used information design very much as statistician and political scientist Edward Tufte has more recently urged, that is, to reveal complexity, to generate knowledge of the world, and to nurture a visual literacy and aesthetic sensibility that is necessary to clear thought, planning, and debate. Without such collaging of planetary systems and trends, projections of 'man's forward evolutionary trending' and the 'ecology of the earth' were unthinkable.

Wigley reminds us that the ecological agenda of 'planetary housekeeping' conceives of the world as architecture. Ecology was, for John McHale and Fuller, 'a certain kind of thinking about or from architecture', Wigley argues. In more detail,

> The strategy is simply to give the world to architecture. The planet is transformed into one big architectural site by the new technologies of communication in the which the world family needs to be rehoused. If the world is a house, it needs designers.[40]

John McHale and Fuller were, first and foremost, image makers, Wigley asserts, for architecture scaled up to the planet. Much of their work with the Inventory, the Geoscope, the WDSD, the World Game, and then the McHales' futurology, consisted of using images architecturally, deploying imaging technologies (such as information design, but also projections, interactive maps, and so on) in order to make architecture itself a plastic and expandable image. Their principal concern was 'a planet rendering problem'.[41] Wigley criticises John McHale for erasing difference, locality, conflict, and violence from these renderings. His planetary ecology transformed the world into a house to be redesigned for a unified human family but doing so neglects 'the politics of the house'.[42] This is a fitting criticism, and one often made of Fuller, the 'planetary homeboy'. However, we should remember that this world architecture was a projection and a pedagogical project, its images open to sometimes absurd futures play.

Playing with Futures

The McHales' futurology itself belongs to a trend that emerged in the late 1950s in several Anglo-American institutions, corporations, and, more slowly, universities. John Williams calls this 'world futures', a

> New mode of ostensibly secular prophecy in which the primary objective was not to foresee *the* future but rather to schematise, in narrative form, a *plurality* of possible futures.[43]

Through the late 1960s and 1970s, world futures moved beyond the study of optimum systems to the study of nonlinear, sometimes chaotic systems. Given the sensitivity of complex systems to initial conditions and their nonlinear functions, long-term forecasts and predictions became impossible. Futurology had to study multiple futures. Notably, increased computational power might allow more variables and possible scenarios into futures modelling, but such models were, once again, inescapably suboptimal. The World Game was an example of world futures, what Williams calls 'a geometric fantasy of epic narrative and global integration' without conclusion.[44] World Game, as discussed in Chapter 1, played with futures, and with the idea of the future as such, especially as it generated what Fuller called the 'Scenario Universe'. Williams states that 'Although he never admitted it, Fuller's own metaphysics precluded the possibility of a World Game winner'.[45]

A stranger character of world futures was shown in 1972, when John McHale and Edwin Schlossberg published *Projex*, a collection of fantastic or absurd projects, or 'selected world situations of varying priorities', under the pseudonyms Emmanuel Lighthanger and Rebus Heaviwait, respectively.[46] The authors describe *Projex* as a 'book to be carried in hand which contains manual descriptions, comments, procedures, outlines, notes and reflections', compiled 'in the interest of widening alternatives and extending choice and chance'.[47] Although it appears as a satire, *Projex* also harbours the more sober ambition of opening our minds to new ways of thinking about the future.

An advertisement for Links Books in the programme for the 1972 Festival of American Folklife describes *Projex* as follows,

> The future is always upon us before we know it. This book, a series of metaphysical fantasies about the evolution of technology and man's control of it, prepares us to deal with the future with our eyes open and our heads on straight.[48]

Projex continues the World Game's displacement of politics to cartography *ad absurdum*. One project proposes to 'bulldoze the earth into an icosahedral form' *à la* Fuller's Dymaxion Projection, although the latter is not named. The surplus earth from this terra-forming could then be used to 'build a spare earth'.[49] The 'Game(s)' section presents three imaginary lands, Future Land, Freudyland, and Marxyland. The authors conjure the first from a list of place names, such as Sea of Planning, Moor of Probability, and Hindsight Hollow. The second and third consist of two parts of a recombined world map. Freudyland condenses Eurasia and Africa north of the equator into a continent, with West oriented to the North (although the compass is paradoxical). Dotted lines divide the Eurasian landmass laterally into 'Erogenous Zone' across the middle and 'Neurotic Zone' below. Sweden is named 'Trauma', the Bay of Bengal is 'Anal Outlet', and 'Id' is found east of Yakutsk. Marxyland similarly condenses the remaining landmasses of the Americas, Greenland, Antarctica, and Australasia, and names regions such as Plains of Surplus Value (the US Mid-West), Cape Antithesis (near Recife, Brazil), and the Inlets of Exploitation (Western Greenland).[50]

Another project, Z7B, makes a more detailed proposal for a 'spare earth':

> As it is obvious that the present earth is very untidy, polluted and generally unmanageable, the proposal is to build a spare earth alongside it. We could then move over temporarily whilst we clean up this one: relocate the rivers, rebuild the cities, re-shape its main form.[51]

Projex takes seriously and literally the view that if 'reality' is constructed, it can be reconstructed and diversified in the future. 'There should be more choice of realities', Lighthanger writes, and *Projex* duly features two proposals for Earth RE-Assembly.[52] Schlossberg and John McHale put world futures into play, seeming to mock the ambition of a scientific futurology. What is more, *Projex* extends the human nervous system and psyche to a planetary scale, as many of John McHale's writings had urged (see, for example, *The Future of the Future*), not through an extended cybernetic system of tele-communications technologies, an informational architecture, but through the politics and the often dark psychology of the house. *Projex* was an anomaly, certainly, but its satire both puts a brake on the project of 'secular prophecy' whilst, strangely enough, educating its readers of the latter's necessity. It also shows collage as a ludic, fantastic, and diversifying conceptual technology for charting the future 'politics of the house'.

Centers for Integrative Studies

SUNY Binghamton opened a new professional school, the School of Advanced Technology (SAT), in September 1967. The following academic year (1968–1969), SAT established two centres, the State Technical Service Center and the Center for Integrative Studies. Its aim was to develop 'new approaches to instruction in computer science, applied mathematics, and general systems theory'. Walter Lowen, the Dean of SAT, predicted it to be 'one of the best damn systems schools in the country'.[53] The CIS offered no formal courses, functioning instead as an 'information gathering, synthesising, and disbursing agency' to study technological trends and to offer guidance on futures planning and research.[54] It was integrative because no one discipline could address such large-scale problems, and it was educational, seeking to introduce transdisciplinary research across the curriculum and to serve as an internationally networked database and information centre.[55]

'Futures research', defined by one reporter on the CIS Binghamton as 'the amorphous field of theorising about the probable and the possible and the necessary in future years', studied the technological, moral, and institutional causes of changes in lifestyle and behaviour.[56] Its aim, John McHale declared, was to scale up social collectivity to the transnational and the planetary, to spread awareness of the 'entire planet as operable life-space', with the latter already the spread of global networks such as telecommunications, intercontinental transport, satellites, and so on.[57] The Center's principal tasks, then, were public pedagogy and curriculum reform, pursued through exhibitions such as *What Is the Future of Life on Earth?* in April 1970 (one and half semesters into CIS Binghamton's operations), which depicted the ecological crisis through photographic panels showing pollution, population growth, and resources. The exhibition was organised by five students of Hinman College at SUNY Binghamton to 'arouse community concern for the environment'.[58] It also reassessed SUNY's purpose and strategy. In early 1970, President of SUNY Stony Brook, John S. Toll, assembled a Panel to 'present alternative programs for the University's future'. The Panel included John McHale along with students, faculty, members of the public, alumni, and industry representatives.[59] Whilst Director of CIS Binghamton, he served as advisor for several extracurricular Innovational Projects Board projects.[60] Again, the CIS Binghamton offered no courses, but provided resources for student research and projects and for the broader public. 'Our client is the public and our goal is education'.[61]

John McHale led a programme in early March 1974 on future technology as part of the Wood Society Programs.[62] Open to students and the public, the programme included a screening of the film *Future Shock*, a 1972 documentary directed by Alex Grasshoff and narrated by Orson Welles after the eponymous book by Alvin Toffler, whose popularity suggested to John a 'widening concern with the future'.[63]

McHale admitted that many problems were too complex for unilateral solutions, meaning responsibility falls to individuals and to nations, rather than to technocrats.[64] Similarly, modelling alternative futures was not an exact science. Complexity could not be evaded but could be addressed through building a network to share knowledge.

Such ambitions were not welcomed by all. The student newspaper, *The Colonial News*, published an unfavourable review of the

CIS and interview with John McHale in March 1970. Journalist Andy Pasztor describes the ambitions of the Center as ambiguous and vague and McHale as awkward, defensive, and suspicious. McHale admits that publicising CIS Binghamton requires some diplomacy, as its premise that social and technological problems cannot be solved through any one discipline might antagonise faculty.[65] A more favourable review appeared four years later in *Pipe Dream*, successor to *The Colonial News*. McHale was optimistic about the future. There was still time to avert the impending technological and environmental crises. 'The situation may be urgent and serious, but it is not hopeless'.[66]

CIS Binghamton produced several reports on student research, seminars, publications, dissemination and networking, and consultations. Despite the Center's recognition, in 1975, it lost its designation as it was absorbed by SAT. The University Research Council (URC) ruled that, first, Centers should 'represent research activities which transcend a single individual's scholarly interests' and, second, Centers should be established only by the President after review by the URC.[67] Hence, the McHales moved to the University of Houston in January 1978, where John became Director and Magda, Senior Research Associate in a new Center for Integrative Studies. The Center joined the UH College of Social Sciences and occupied an office on the fifth floor of the MD Anderson Memorial Library. UH Vice President and Dean of Faculties Barry Munitz hailed the CIS as 'one of the most important activities on the central campus' of UH.[68] As before, the aims of the CIS were to study the future implications of current and projected social and technological changes. Integration was required to address the global complexity and consequence of such changes. 'No problem in this complex society can be handled independently', McHale declared in 1978.[69] Both expressed optimism that, by the late 1970s, a 'world commitment' to such problems had begun to consolidate. International collaboration was becoming more common and there was a broad network of initiatives in interdisciplinary research and world modelling, of which the CIS was one of the primary nodes (Figure 4.3).[70] Again, both asserted that the problem was not scarcity of resources but their distribution. Shortages were a distributional problem. 'We don't have an energy shortage. We have a problem with overuse of some fossil fuels'.[71] Maldistribution and the lack of political will were also the causes of ongoing poverty.[72]

John McHale died of a heart attack on Thursday 2 November 1978 at Twelve Oaks Hospital, Houston. Following his death,

Figure 4.3 Aurelio Peccei receiving the key to Houston from the McHales;
   University of Houston Libraries Special Collections/1969-037,
   UH Photographs Collection/People (Groups), Box 42, Folder
   44. Courtesy of University of Houston Special Collections UH
   Photographs Collection.

Magda McHale continued her futures studies with the publication
of their global study (co-written with Guy Streatfield) on the needs
of children, *Children in the World*, commissioned by the US Population Reference Bureau to coincide with the UN International
Year of the Child.[73] Magda explained that their recommendation
was to raise children to 'some minimum standard of food, sani-
tation, health and education'.[74] Despite the rather dire situation
faced by many children, Magda remained optimistic. 'I am very
encouraged that human beings are infinitely resourceful and that
basically humanity as a whole has some kind of integrity'.[75] *Children in the World* was disseminated worldwide as part of a package
also containing a special edition of the Bureau's *Population Bulletin*
titled The World of Children and a chart, the *1979 World's Children
Data Sheet*, showing demographic information for children from
154 countries.[76]

In 1980, Magda moved to the University of Buffalo to set up a CIS in the Department of Environmental Design and Planning in the School of Architecture and Environmental Design. The Executive Vice President of SUNY Buffalo and Professor of Political Science, Albert Somit, wrote enthusiastically in support of McHale's appointment to a research professorship and congratulated the School on attracting her. He noted the McHales' 'positions of pre-eminence in planning, futurism, and resource and technology management'.[77] Harold L. Cohen, Dean of the School and long-time friend of the McHales, spoke with even more enthusiasm about the uniqueness of the CIS as a data resource – '57 filing cabinets of research data and 8000 volumes in the fields of architectural design, world resources planning, world ecology, new towns and cities, and population planning' – for students of architecture and design, policy studies, management, and law. McHale, a 'one-woman dynamo', would also serve as a role model for at least 30% of the School's student body.[78] The CIS would provide the School with a 'window on the world'.[79]

Among Magda McHale's major projects during the first years of her tenure were the conferences *Approaches to the Study of the Future: United States, Canada, and Mexico* in May 1980 and *Information Revolution: Different Socio-Cultural Perceptions* in April 1983, the latter sponsored by the Intergovernmental Bureau for Informatics, Rome. One consideration of the *Information Revolution* conference, attended by twenty people from nine countries, was the instability and complexity that resulted from the incongruence of the changing 'ecology of knowledge' with cultures' ability to adapt. The conference group tested scenarios of the impending 'information revolution', based upon models ranging from 'the present predominant model, mainly driven by the profit motive and military expenditures, to a model inspired by the approaches and attitudes of women (sharing and nurturing)'. Each model included three types of networks, industry and technoscience, academia and other knowledge communities, and those communities and societies still 'struggling to design their own development' but often 'marked by a greater degree of caring and sharing'.[80] The group recommended that 'those involved in the development of informatics, especially in the industrialised world, should recognise the importance of helping cultures to adapt to the revolutions in progress'.[81] The previous year, Magda McHale wrote *Reflections on Impacts of the Information Environment*, in which she already advocated sharing of the 'requisite information and communications systems for social navigation'.[82]

From the preceding, we see that the McHales' futurology depended upon the promotion of 'integrative studies'. This latter continued their project of 'cultural collage', of network building, which was far from unitary or homogenising. Their institutional collage was necessary to track the decentralising trend of the state form in the short term and the emergence of a transnational 'world community' in the longer term.[83] Similarly, they required an integrative, informational medium, which was, first and foremost, the diagram.

It is through these two – a cultural collage of institutions and a visual collage of information/the diagram – that the McHales' project of a planetary ecology was expressed. Certainly, planetary ecology was a thoroughly humanist project, at times paternalistic and technophilic, but it was also a patching together of dissimilar elements, a collage and a poetic metaphor, whose absurdity McHale seemed to acknowledge (and play with) on at least one occasion. This would suggest that, to return to Wigley's metaphor, there was some attempt here to address the internal politics of the house.[84]

Notes

1 Magda Cordell McHale in Gay E. McFarland, 'The Future: Couple Share a Job of Peering at Years Ahead', *Houston Chronicle*, Sunday 11 June 1978, 20.
2 John McHale and Magda Cordell McHale, 'World Trends and Alternative Futures', *Open Grants Papers* no. 1 (Honolulu: East-West Center, 1974), 1.
3 McHale and McHale, 'World Trends and Alternative Futures', 39.
4 McHale, 'World Dwelling', 164.
5 McFarland, 'The Future', 20.
6 Stanford University Collections M1090, Series 18: Project Files, World Game Subseries 2, box 30, Folder 4: World Game History 2, Personnel 1966–1968 (2 of 2).
7 Stanford University Collections M1090, Series 18: Project Files, World Game Subseries 2, box 24, Folder 15: Progress Report (Carl G. Nelson) April 1 1969.
8 Anon. 'SUAB Director Authors Book on Ecology', *Vestal News*, 24 November 1970.
9 John McHale biography, Binghamton University Archives, Special Collections, Binghamton University Libraries, Binghamton University, State University of New York.
10 Bebe Landry, *SUNY Binghamton News*, 9 September 1971, 30.
11 McHale obituary, Center for Integrative Studies, University of Houston, November 1978, Binghamton University Archives, Special Collections, Binghamton University Libraries, Binghamton University, State University of New York.

12 John McHale quoted in Harold A. Linstone, 'Editorial Comment: The Rome Special World Conference on Futures Research', *Technological Forecasting and Social Change* 5, no. 4 (1973), 329a.

13 Bebe Landry, *SUNY Binghamton News*, 4 September 1973, 30.

14 Anon. 'SUNY Gets $20,000 UN Grant', *The Sun Bulletin*, 19 February 1974.

15 Magda Cordell McHale and John McHale, 'Future Studies: An International Survey', *Ekistics* 41, no 246 (May 1976), 300–306. Magda Cordell McHale and John McHale, with Guy Streatfield and Laurence Tobias, eds. *The Futures Directory: An International Listing and Description of Organizations and Individuals Active in Futures Studies and Long-Range Planning* (Guildford: IPC Science and Technology Press, 1977).

16 Bebe Landry, *SUNY Binghamton News*, 11 September 1974, 30.

17 John McHale, Magda Cordell McHale, and Guy Streatfield, *Women in World Terms: Facts and Trends* (Houston, TX: University of Houston, Center for Integrative Studies, 1975).

18 See Giulia Smith, 'Painting that Grows Back: Futures Past and the Ur-feminist Art of Magda Cordell McHale, 1955–1961', *British Art Studies* 1 (2015).doi:10.17658/issn.2058-5462/issue-01/gsmith.

19 John McHale, *The Ecological Context* (New York: George Braziller, 1970), 1, 166.

20 John McHale, 'The Plastic Parthenon', in *The Expendable Reader*, ed. Alex Kitnick (New York: GSAPP Books, 2011), 97.

21 John McHale, 'The Future of Art and Mass Culture', in *The Expendable Reader*, ed. Alex. Kitnick (New York: GSAPP Books, 2011), 246.

22 McHale, *The Ecological Context*, 3.

23 McHale and McHale, 'World Trends and Alternative Futures', Abstract.

24 Magda Cordell McHale and John McHale, 'Organisations-Methods Index', in *The Futures Directory: An International Listing and Description of Organizations and Individuals Active in Futures Studies and Long-range Planning*, eds. John and Magda McHale (Guildford: IPC Science and Technology Press, 1977).

25 John McHale, Magda Cordell McHale, and Guy Streatfield, *Women in World Terms: Facts and Trends* (Houston, TX: University of Houston, Center for Integrative Studies, 1975), 2.

26 Ben Highmore, *The Art of Brutalism: Rescuing Hope from Catastrophe in 1950s Britain* (New Haven, CT: Yale University Press, 2017), 263.

27 Mark Wigley, 'Never at Home', in *The Expendable Reader*, ed. Alex. Kitnick (New York: GSAPP Books, 2011), 283.

28 Anthony Vidler, 'Whatever Happened to Ecology? John McHale and the Bucky Fuller Revival', *Log* 13/14 (Fall 2008), 146.

29 John McHale, *Fathers of Pop* (1979), discussion between Mary and Reyner Banham and Magda Cordell McHale and John McHale, 29–30.

30 Quoted in David Robbins, ed. *The Independent Group: Postwar Britain and the Aesthetics of Plenty*, exhibition catalogue (London: Institute of Contemporary Arts, 1990), 87.

31 This is the claim made by Juliette Bessette in 'John McHale: de l'art du collage à la pensée prospective', *Les Cahiers du Mnam* 140 (Summer 2017), 40.

32 John Bender and Michael Marriman, *The Culture of Diagram* (Stanford, CA: Stanford University Press, 2010), 35.
33 Edward R. Tufte, *Envisioning Information* (Cheshire, CT: Graphics Press, 1990).
34 John McHale Statement, Private view card for exhibition '3 Collagists - new work by E L T Mesens, John McHale, Gwyther Irwin' (recto), Institute of Contemporary Arts, London, 5–29 November 1958.
35 McHale, 'The Future and the Functions of Art', 193.
36 John McHale, *Phase II (1967), Document 6, The Ecological Context: Energy and Materials* (Carbondale, IL: World Resources Inventory, 1967), 59.
37 McHale, *Phase II (1967), Document 6, The Ecological Context*, 24.
38 McHale, *The Ecological Context*, 174, 4.
39 John McHale, 'Global Ecology: Toward the Planetary Society', *The American Behavioral Scientist* 11, no. 6 (July/August 1968), 29.
40 Mark Wigley, 'Recycling Recycling', *Interstices: Journal of Architecture and Related Arts* 4 (2019), 5, https://interstices.ac.nz/index.php/Interstices/article/view/589.
41 Wigley, 'Recycling Recycling', 11.
42 Wigley, 'Recycling Recycling', 12.
43 R. John Williams, 'World Futures', *Critical Inquiry* 42 (Spring 2016), 473.
44 Williams, 'World Futures', 503.
45 Williams, 'World Futures', 508.
46 Rebus Heaviwait and Emmanuel Lighthanger, Foreword to *Projex* (New York: Links Books, 1972), unpaginated.
47 Heaviwait and Lighthanger, Foreword to *Projex*.
48 Links Books advertisement in Gerald Davis and Ralph Rinzler, Programme for 1972 *Festival of American Folklife* (New York: Music Sales Corporation, 1972), 49.
49 Heaviwait and Lighthanger, *Projex*, 30.
50 Heaviwait and Lighthanger, *Projex*, 72–75.
51 Heaviwait and Lighthanger, *Projex*, 102.
52 Heaviwait and Lighthanger, *Projex*, 30.
53 Quoted in Lawrence A. Moss, 'Sat: A Small Glimpse of the Future Today', *The Colonial News*, Friday 13 March 1970, 4.
54 Moss, 'Sat: A Small Glimpse of the Future Today', 4.
55 Anon./Center for Integrative Studies, Brief Description, Binghamton University Archives, Special Collections, Binghamton University Libraries, Binghamton University, State University of New York.
56 Anon, 'Research in the Present about the Future', *Colonial News*, Tuesday 5 March 1970, 5.
57 McHale, quoted in 'Research in the Present about the Future', 5.
58 Anonymous press release, *The Colonial News*, Thursday 14 April 1970, 2.
59 Barbara Ray, 'Panel to Investigate University Purposes', *The Colonial News*, Friday 13 February 1970, 2.
60 Peter Salgo, 'Innovational Projects Board', *Pipe Dream*, Friday 23 April 1971, 4.
61 McHale, quoted in Andy Pasztor, 'A Center to study Implications of Technological Developments', *The Colonial News*, Friday 13 March 19706.

62 Anon. 'The Woods Society', *Pipe Dream*, Friday 30 November 1973, 11.
63 John McHale, 'The Changing Patterns of Futures Research in the USA', *Futures* 5, no. 3 (June 1973), 262.
64 McHale quoted in Richard Schroeder, 'Woods Society—McHale and the Future', *Pie Dream*, Friday 8 March 1974, 21.
65 Pasztor, 'A Center to Study Implications of Technological Developments', 6.
66 McHale, quoted in Robert Lowe, 'Future Tense or Future Perfect?' *Pipe Dream*, Friday 22 November 1974, 21.
67 Letter from John E. La Tourette, Provost for Graduate Studies and Research, to John McHale, 10 November 1975, Binghamton University Archives, Special Collections, Binghamton University Libraries, Binghamton University, State University of New York.
68 Jim Asker, 'UH Think Tank Involved in Study of World's Ills', *The Houston Post*, Sunday 5 June 1977, 8B.
69 Quoted in McFarland, 'The Future, 20.
70 McFarland, 'The Future', 20.
71 McHale in Asker, 'UH Think Tank', 8B.
72 Ellen Perlmutter, 'Poverty Can Be Erased, Sociologist Says', *The Sun Bulletin*, 14 January 1976.
73 Magda Cordell McHale and John McHale, *Children in the World* (Washington, DC: Population Reference Bureau, 1979).
74 Quoted in Bill Coulter, 'Millions of Children Face Bleak Future, UH Study Finds', *The Houston Post*, Sunday 11 March 1979, 3D.
75 Quoted in Coulter, 'Millions of Children Face Bleak Future', 3D.
76 Wendy Adair, 'International Year of the Child Hailed by New UHCC Global Report', *UH Central*, 5 January 1979.
77 Albert Somit, letter to Harold L. Cohen, dated 5 November 1979.
78 'Cohen 'Ecstatic' over Attracting Major Center', *Reporter* 11, no. 3 (20 September 1979), 2.
79 Joe Simon, 'Center's Move North Gives UB School "Window on the World"', *The Spectrum*, Wednesday 19 September 1979, 3.
80 Magda Cordell McHale, *Summary Conference Report of the International Conference on Information Revolution: Different Socio-Cultural Perceptions*, April 1983, University of Buffalo Special Collections, Center For Integrative Studies 1982–1983, Folder 1, 4.
81 McHale, *Summary Conference Report*, 6.
82 Magda Cordell McHale, *Reflections on Impacts of the Information Environment*, prepared for the Intergovernmental Bureau for Informatics, Rome, 1 June 1982, University of Buffalo Special Collections, Center for Integrative Studies 1982–1983, Folder 1, 2.
83 McHale and McHale, 'World Trends and Alternative Futures', 35.
84 Mark Wigley, 'Recycling Recycling', *Interstices: Journal of Architecture and Related Arts* 4 (2019). https://interstices.ac.nz/index.php/Interstices/article/view/589.

5 World Gaming

Howard Brown and Medard Gabel founded the World Game Institute (WGI) in 1972 as a non-profit corporation to promote the World Game through workshops, seminars, and educational resources. WGI developed the World Game Workshop (WGW) in 1986, and in 1990, began to construct a 'world brain' saved on computer disks. The first version of the latter, Global Data Manager, was a spreadsheet of global data sets, described by Gabel as the 'largest database of socioeconomic and environmental indicators for the world', featuring up to 15,000 statistics per country. Global Recall 2.0 followed in 1993, a package including over 500 maps, 600 statistical indicators for each country, an encyclopaedia of world problems, and a Solutions Lab where users are trained in problem-solving and are tasked to devise strategies to meet these problems. The Lab includes an econometric model of nine sectors such as Food, Education, Energy, Environment, so that users can manipulate variables.[1] They then can enter their solutions into the World Game Tournament. Global Recall, Gabel claims, democratises access to global information and 'legitimises the study of complex problems by the lay person'.[2] Since 2019, the WGW and its databases have been in the care of the Schumacher Center for a New Economics. Through these and other versions, the ambitions and protocols of the World Game have continued but have also adapted to changes in technology and to serve different clienteles and publics. Just as often as children have been its primary players, the World Game has been pitched at corporate clients and 'anticipatory leaders'.

This final chapter surveys the World Game's evolution up to the present and compares it to other pedagogical experiments in world gaming, all of which abandoned classroom hierarchies and promoted a constructivist pedagogy of 'learning to learn', where the aim was for players to acquire the flexibility and confidence to adapt critical skills collectively to social, ecological, and, at times, political problems. Crucially, however, rather than focus only on the learning subject, these world gaming experiments sought to orient play's fun, flow, and conviviality towards what Gert Biesta has called a 'world-centred education'.[3] What is more, the project of 'making the world work', still most often presented as a design and engineering problem, became, sometimes explicitly, an exercise in political economy. This returns us to an ongoing dispute over the nature of world gaming and planetary management: who does it, for what purpose, at what scale, and by what means? We return, that is, to that issue raised by an audience member at the first World Game Seminar, namely, the politics of implementation.

Willard Junior High School

The first example took place a year before the World Game Seminar at New York Studio School of Painting and Sculpture when Edwin Schlossberg and architect Jon Dieges ran an architecture and urban planning course at Willard Junior High School in Berkeley, California. Approximately thirty students aged thirteen and fourteen met for two hours a day, five days a week, from 20 June to 19 July. The single-sheet *Good News: Journal of Environmental Design* described this course's flexible curriculum. Students were asked what they would like to build (Figure 5.1). Their comments printed on the recto laud this flexibility and freedom above all else.

> [The students] designed space capsules, a city, localised communication or circulation areas, games, a new plan for education, a new way to distribute population through the world, a new building complex for education.[4]

They were introduced to Fuller's 'synergetic' geometry. One photograph shows students building a geodesic frame. Another shows a student holding to the camera two icosahedral Dymaxion maps. A third shows two students discussing with Dieges how to build a geodesic using a globe map as support. They were asked about their local environment and how this was affected by ecological changes and were encouraged to think of themselves as 'energy systems', like the metabolic selves of the World Game.

A week into the Willard course, on Sunday 30 June, Berkeley mayor Wallace Johnson declared a state of emergency and a three-day curfew after two nights of confrontations, barricades, and fires by up to 4,000 protesters in support of the strikes in Paris.[5]

Figure 5.1 Photograph of course given by Schlossberg and architect Jon
Dieges at Willard Junior High School, Berkeley, CA from 20
June to 19 July 1968, M1090, Series 18: Project Files, World
Game Subseries 2, Tom Turner Files, box 105, folder 2. Cour-
tesy of the Estate of R. Buckminster Fuller.

These events prompted the Willard students to consider destroy-
ing their city. 'Many of the students saw the violence, saw the de-
structive nature of their contemporaries and they came to school
on Monday with several ideas on how to destroy their model city'.[6]

Despite this, they completed their model, played music, recorded a play, and taped conversations with their teachers Ed and Jon.

The Willard course evidently forced players to engage with governmental and ecological politics, especially the concrete conditions of protest and policing in Berkeley at that time. Their desire to destroy their model city shows the students' awareness of the correlation between the built environment and political order. The Willard course also belongs with several similar pedagogical initiatives in Berkeley around this time, which also taught children and students an open curriculum of geodesics, recycling, environmental design, and self-governance.[7] The following year, Schlossberg co-organised the World Game Seminar. The Willard course shows the World Game's connection to pedagogy of children, especially through extra-curricular seminars and workshops in participatory design and planning. In some respects, the World Game Seminar upscaled the Willard course to planetary management.

By the early 1970s in US schools, 'educational gaming' had become a movement of sorts. Many of its simulation games promoted environmental education, combining interdisciplinary knowledge in 'operating models' of real-world problems and situations. Such games, introduced from elementary to high school and addressing land use, pollution, waste and recycling, air quality, and urban growth, provided pupils with the opportunity 'to participate in environmental citizenship'.[8] World gaming belongs to this trend, upscaled to planetary management and geopolitics.

John Hunter's World Peace Game

One renowned example of this upscaling, not directly connected to the World Game and its derivatives but close to these latter in its design and aims, is the World Peace Game (WPG) devised by fourth-grade teacher John Hunter in Charlottesville, Virginia, in 1978 and still practiced today. Hunter describes the WPG as a 'geopolitical simulation' and 'a multi-dimensional matrix that contains a world ecosystem riddled with a host of situations and issues in flux and at play'.[9] Its broad aim is to prepare schoolchildren for the responsibility of solving global problems collectively and promoting peace.

The game board for the WPG consists of a stack of four plexiglass layers that support the undersea level, ground and sea level, air and space level, and outer space level. Around the board, there are four countries, each having a profile, budget, military and commercial assets, and a rudimentary cabinet (Figure 5.2). Hunter's notes on an initial sketch show the WPG as an exercise in realpolitik played

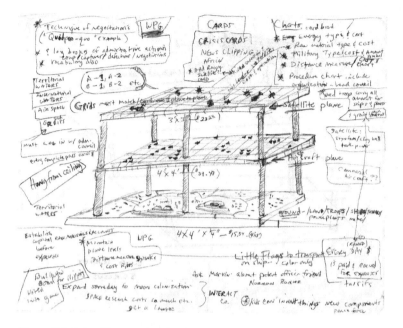

Figure 5.2 John Hunter, sketch of World Peace Game, 1978. Courtesy of World Peace Game Foundation.

out on a three-dimensional gameboard. Prime ministers are chosen by the teacher overseeing the WPG. There is also a World Bank, a United Nations, arms dealers, and a weather goddess who controls the stock market, weather, and climate, and can have a decisive role in resolving issues.

The players' primary challenges are outlined in a document that details fifty correlated crises (based on real-world events) that include environmental disasters, religious and ethnic conflicts, famine, nuclear proliferation, separatism, and biodiversity loss. Thus, WPG throws its players into a situation of geopolitical complexity, contradiction, and near-crisis that is not, initially, of their own making.

> The ultimate challenge is for students to understand through game play that if one thing changes, everything else changes because problems are complex, interdependent, and far-reaching in their consequences.

The game, which typically takes 16 to 20 hours of class time and is played over two to three months (or during an intensive one-week course in summer), is won when every nation's net asset value has risen beyond its starting point and all game crises have been solved.[10]

What do students learn? Role play, problem-solving, 'record keeping', critical thinking, 'speculation (based on logical and intuitive inferences)', diplomacy, compassion, and tolerance, but also, presumably, a sense of political strategy and of dealing with unpredictable events across a global arena.[11] Students would also learn elements of futurology. 'The student will analyse how future problems emerge from current trends, extrapolate problems from trends, and evaluate and negotiate for, solutions to those problems'.[12]

To introduce confusion and disruption, there is a 'covert saboteur' who seeks to undermine the WPG through 'misinformation, ambiguities, and irrelevancies'. This agent (often one of the brightest students of the class) prompts unforeseen changes to the game situation, which, Hunter argues, pushes all players towards deeper critical thinking. Players are challenged to identify this saboteur through a brief trial. If they are wrong, the accusers must pay a fine. The saboteur, along with other uncontrollable variables, produces a refracted view of the world, a fiction that produces real crises. These produce VUCA, or volatility, uncertainty, complexity, and ambiguity, to use the acronym coined by management theorists Warren Bennis and Burt Nanus in 1985, and, Hunter claims, promote collectivity and leadership in response.[13] Victory is defined by all the crises being resolved and every nation increasing its asset value beyond its starting point.[14]

Hunter conceived of WPG in 1978 for a social studies class at Richmond Community High School to teach about African current affairs. His choice of an amended game board to teach came from his mentor Ethel Banks' advice to 'build curriculum around what [students] love'.[15] The WPG conceives of the classroom as an empty space, one that allows players flexibility, openness, and the possibility of not acting until the right moment. This learning space is not filled with assessment and outcomes ('absorbing knowledge and demonstrating results'),[16] and the aim is not for students to arrive at a correct answer (known in advance by the teacher) but to organise themselves into a collective that can devise and implement negotiated solutions. Hunter answers in one of four ways to his students' questions: What do you want to do? Can you afford it? Have you considered the consequences? Does it make sense? In this, he trusts to the students' intelligence and ingenuity and to the intelligence and 'collective wisdom' of the game itself to guide play.

There is a Taoist influence upon this empty space. As students play WPG, Hunter reads to them from Sun Tzu's *The Art of War* so that they might understand war and thereby avoid it or, if war is unavoidable, exit it as rapidly as possible, but also so that they learn the importance of being flexible and waiting to act. They might also learn to understand that appearances often belie deeper truths.[17] The WPG is designed 'to allow space for a deep, extended, continuous review of the students' ideas and thoughts', a type of slow learning.[18] Hunter describes the WPG as 'a vehicle for stimulating higher-level critical and creative thinking, but also as a laboratory for eliciting the best in human intentions', namely, compassion, compromise, collaboration.[19] Players self-assess their performance using a set of 'thermometers', graded 1 to 10, for function, elegance, effort, and resources used. They are also asked to list the intelligences they have used (according to Howard Gardner's taxonomy) and to fill an empty with a map of their thought process.[20] Once completed, their self-assigned grades, or 'temperatures', are then negotiated with Hunter.

Broadly speaking, WPG uses the same pedagogy as other world gaming experiments. Yet, it introduces planetary management to the messy and complex world of geopolitics. What is more, in the figure of the saboteur and the random behaviour of weather and market, players encounter Black Boxes that are not defined only by an absence of information (or data). They are instead agents of disruption, even malign intent, and they are, by design, unpredictable.

WPG has its biases, of course. The measure of success is akin to GDP rather than, say, quality of life, paramilitary groups seem not to be open to negotiation, it presents markets as a force of nature akin to weather and neglects the wickedness of certain problems, especially environmental problems. Nevertheless, to repeat, it promotes world gaming as a geopolitical simulation rather than a simulation of design or engineering problems. Most importantly, it allows game rules to evolve under pressure from player demands, which rarely occurs in versions of the World Game, expect when produced for corporate clients, as discussed below.

GENI

A more direct descendant of the World Game, which, by contrast to the first two examples, restricts world gaming to engineering problems, is the Global Energy Network International (GENI), founded in 1986 by engineer Peter Meisen after he was advised by Fuller to read the latter's *Critical Path*. GENI follows Fuller's demand for a global electric energy grid by developing a model to show the benefits of how to connect remote renewable energy sources to populations across the globe. 'This global option', GENI claimed, 'is the highest priority solution from the World Game'.[21] Fuller's *Critical Path*, published in 1981, repeated his assertion that raising living standards for all humanity within ten years was 'not an opinion or a hope – it is an engineeringly demonstrable fact. This can be done using only the already proven technology and with the already mined, refined, and in-recirculating physical resources'.[22] That the World Game had proven the feasibility of its design solutions was a frequent claim of its advocates during the World Game Seminar and after. Similarly, much of GENI's work through the 1990s involved establishing their model as both feasible – 'an engineeringly demonstrable fact' – and desirable through workshops, conferences, broadcast media, exhibitions, and, in April 1996, two WGWs in San Diego and Santa Barbara.[23] GENI employs Fuller's notion of 'comprehensive anticipatory design science' and the model uses the Dymaxion Projection to show this global transmission network. It follows Fuller's assertion that the primary driver of living standards is kWh per capita. Beyond a threshold of 2,000 kWh per capita p.a., life expectancy, literacy, and access to safe drinking water all increase and infant mortality decreases. For comparison, the US average in 1987 was over 11,000 kWh per capita p.a. The authors note (in 1995) that although their software struggled to model many variables across different time frames, the main obstacles to implementing GENI were financial, social, and political.[24]

In many respects, GENI continues the most technocratic and technophilic aspects of the World Game, such as when its Global Model Index,[25] a history of global simulations, is written largely as an evolution of programming languages. In this respect, GENI continues the depoliticising tendency of the early World Game, coupled with Fuller's belief in computational power as the ultimate tool to improve design science solutions. Mark Wasiuta notes, however, that GENI and other examples from the Global Model Index remain limited by what a model can encode. By contrast, Fuller sought to 'rewrite the world *as* code', producing, so to speak, a map scaled up to its territory.[26]

GENI also produced a World Resources SIM Center on the model of the unbuilt World Resources Simulation Center. First prototyped in 2009 and established as the SIM Center in San

Diego in 2011, its aim, broadly conceived, is to 'visualise sustainable solutions' in an immersive environment to display critical issues and trends and to simulate future scenarios. The Dymaxion Projection takes centre stage as the game board. For the prototype in June 2009, teams of GIS experts and 'comprehensive thinkers' (entrepreneurs, scientists, etc.) addressed issues of population demographics, energy and climate, and access to water across scales from local to international. The SIM Center continues as one of the arenas for WGWs.

World Game Workshops

As noted, the WGI began its Workshops in 1986. Through the 1990s, the Workshops became the centrepiece of the WGI's activities, played by around 250,000 players in thirty-five countries, with clients across the educational, non-profit, and corporate sectors (Figure 5.3). These were team games played barefoot or in stocking feet on a large Dymaxion Projection spread across the floor of a hall, accompanied by slide and acetate projections, gamecards, a currency (in US Dollars), folios of regional problems, information desks, fact sheets detailing human needs, the World Game Handbook, and Global Recall 2.0 as a learning tool for before and after the Workshop. The general aim of the Workshop was to simulate a global economy as a game board for strategy training. 'We are going to put you in charge of the World', facilitators would declare at the start. Its Corporate Program offered a 'four-hour experiential program' for conference networking or as a training module in global leadership, teamwork, and corporate citizenship, adaptable to the client's needs.[27]

The Workshop would have three parts. First, a Global Overview, which would introduce players to the planet as an integrated system, to a sense of deep time, and to the scale of the map through a sequence of demonstrations and slide projections. Second, a World Simulation, in which players were to deal with 'complex, multidimensional, non-linear real-world problems' through applied systems thinking and team action. In the third part, Alternate Futures, teams are requested to envision an 'unacceptable future' that might prompt greater social and corporate responsibility.

In the version for students and teachers, there would be eleven regional teams and six to ten international teams such as the WHO, UN, UNESCO, and World Bank.[28] Teams were asked to address problems of climate change, biodiversity loss, loss

Figure 5.3 World Game Workshop, American Forum, Indianapolis Statehouse, June 1993. Courtesy of Schumacher Center for a New Economics Library. Photograph: Lisa Mandell.

of habitat, pollution, and human population growth. After the Workshop, students would be encouraged to pursue careers as 'problem solvers who have a vision of a better future' and could read through a two-page list of occupations that might end deforestation, provide healthcare, manage and retire debts, or provide education.[29] In the corporate version, these teams would be joined by corporations, media networks, and the Invisible Hand and a Production Center staffed by facilitators to introduce 'shifts and uncertain market conditions' and to transform raw materials into products, respectively.[30] There is also a Marketboard that sets initial prices.

In these Workshops, the World Game integrates the geopolitics and realpolitik largely lacking from the Seminar, as teams of players, wearing paper hats or name badges to show their affiliation (functioning as 'passports'),[31] negotiate with other regions and with international institutions. In this, it brings the World Game closer to WPG. In the corporate version, especially, and again like the WPG, it also naturalises a market mechanism as a background variable within the game world, something to which players must react but over which they exert little control unless a client wishes that the Invisible Hand 'reflect some of the issues facing the company'.[32]

The WGW's success led Brown to found the for-profit o.s.Earth (operating systems Earth, OSE) in 2000. OSE's 'flagship product' was The Global Simulation, a live-action game where, again, players take on the role of global leaders.[33] A promotional video shows The Global Simulation being played by twenty teams representing regions, with corporations competing for profit, organisations developing and selling solution strategies, all observed and assisted by teachers and administrators. Although available for schools and colleges to hire, in this version, its gameplay reflects further the needs of its corporate clientele, with rounds of resource competition, trading, and price setting.

Following a similar path to Brown, Gabel led BigPicture Consulting from 2003 to 2016, which ran several multimedia interactive simulations – FutureGame, Green Company, and Innovation Game – available as tools for globalisation strategy. FutureGame, for example,

> is used for mid- to long-range strategic planning, envisioning realistic future goals, opportunities and challenges, new product development and innovation opportunities, as well as team building in a context of organizationally relevant problem solving.[34]

Clients include Motorola, Burger King, General Motors (GM), International Business Machines (IBM), Novartis, OSI, Delta Airlines, British Airways, MasterCard International, and Chase Manhattan Bank. Gabel and Jim Walker promote Fuller's work as a model of 'anticipatory leadership', or more precisely, a 'problem-solving leadership framework ... useful not only for changing the world, but also for changing your own local organisation of business – an equally challenging task'.[35] For Gabel and Walker, 'anticipatory design science' is in the business of forecasting and optimal solutions, anticipating trends so as to be neither too early nor too late to market. The FutureGame builds upon this corporate version of design science to provide clients with a tool to help globalise their company. Its principal resource is *Global Inc. An Atlas of the Multinational Corporation*, written by Gabel and Henry Bruner and published in 2003, which offers clients a history and conceptualisation of multinational corporations up to the present accompanied by over 400 maps, charts, and graphs. *Global Inc.* was funded by Ford Foundation and includes an introduction by John Browne, then CEO of British Petroleum, who lauds economic globalisation for making the world work for everyone. Writing in the *Wall Street Journal*, George Melloan praised *Global Inc.* for 'feeling the muscles of the multinationals' as the latter triumph over the anti-globalisation protests at Seattle and elsewhere.[36] Continuing that metaphor, BigPicture Consulting makes the World Game HIIT for those multinationals, a further example of world futures gaming for the business sector.

Through these versions, the World Game, along with the Workshops in which it was played and the design science upon which it was founded, was realigned with a corporate rationale. Design science's ambition to 'do more with less' would offer no challenge to profit margins and a sustainable future would be one that sustains a market economy and private accumulation. In such a case, the global perspective of world gaming seems indistinguishable from the current 'world-system' identified by Immanuel Wallerstein, that is, the globalised economy with its structural inequalities and externalities, its technologies of extraction and accumulation, managed by large-scale corporate agents that determine proper or sustainable use.[37] It would also assume that current crises require little socioeconomic or political change, only improved design and new markets.

In 2019, however, the Schumacher Center for a New Economics bought the rights to the WGW. Founded in 1980, the Schumacher Center advocates local and regional economies, community revitalisation, and decentralised networks in keeping with the economist E. F. Schumacher's distrust of technological 'gigantism', standardisation, and large-scale, centralised management of the environment and its resources. It is intriguing, then, that the Schumacher Center would choose to enlist a tool for planetary management, the WGW, to its Civic Synergy Program to promote 'the collective power of people in organized networks to transform the systems that affect their lives'.[38] In fact, led by Greg Watson, Director of Policy and

Systems Design at the Center, and Elizabeth Thompson, former Executive Director of the Buckminster Fuller Institute, the WGW aims to teach global systems literacy in an institution committed to localism, what Fuller once called 'local-focus hocus pocus' just as John McHale declared, two years after Fuller, 'There are no "local" problems anymore'.[39]

With the support of the Institute of Electrical and Electronics Engineers and the SIM Center, the WGW has proposed a global high-voltage transmission network (following GENI) to accelerate renewable energy uptake. This is one example of an 'accelerator' to promote speedy, multi-scalar, international optimal design to address the crises we face. The WGW itself provides a 'mission control' for global sustainability in which to visualise trends and futures, immerse players, enable collaborative problem-solving, and connect researchers, engineers, economists, et al. 'to design, fund, and build solutions quicker'.[40]

A sense of WGW gameplay is provided by the Global Emergency Production Board, on the model of Franklin D. Roosevelt's War Production Board, formed in January 1942. This seeks to mobilise the economy towards a serious of global threats (climate crisis, nuclear war, wealth inequality, biodiversity loss). As with previous versions, materials include a large acetate Dymaxion floor map; a player's manual; cards for world resources, needs and trends, systems literacy, technology; resource tokens; green infrastructure nodes; and political currency. Whether it might be accurate or productive to treat the climate crisis as a war, moving the WGW to a war footing indicates both the gravity of the threats it now models and the perceived necessity of state intervention to repurpose the economy.

Considering this, the ongoing redesign of the WGW's materials, technologies, and gameplay seems especially timely. Updates include using the En-ROADS climate policy simulation and indexing the game time of WGW to temperature and emissions targets.[41] Another update will be to model the extraction technologies and supply chains for renewable energy sources, especially concerning the extraction of rare earth elements and the embodied energy of sustainable technologies and infrastructures. Finally, there is the possibility of adapting the WGW to those remote sensing networks that now, as Jennifer Gabrys notes, make the planet and its environments computable and programmable. Through monitoring wildlife populations, pollutants, and other climatic variables, these technologies advance the 'becoming environmental of computation' but also produce new practices of sensing and operationalise data 'into new forms of environmental engagement and experience'.[42] These 'citizen sensing practices', Gabrys argues, do not necessarily reduce the democratic deficit in planetary management but offer contemporary expressions of environmental problems, their agents, and their politics.[43]

It is intriguing how an updated WGW might offer a trans-scalar architecture that correlates planetary management and design science with the 'cosmo-local' and pluriversal or that promotes a future of global design matched to local manufacturing and local economies.[44] This would return the World Game to 'technoanarchy' and to the futurology of the McHales' Centers for Integrative Studies through the WGW's combination of citizen science, environmental computation, real-time data, and the commitment to 1:1 planetary modelling.

Whatever updates Watson and Thompson might introduce, they are done in the understanding that the World Game is a tool for testing solutions, certainly, but also for defining and modelling new problems, its aim being to show what can be done at a given moment.[45] It therefore remains a 'mountain-climbing game' without a peak, but at the Schumacher Center, the WGW might also become a 'mountain-climbing game' for commoners as it brings global systems literacy together with projects that 'reinvigorate democratic self-governance'.[46] Might the WGW become one of those 'insurgent projects' described by David Bollier, Director of the Reinventing the Commons Program at the Center?

> They reject the standard ideals of economic development and a return on shareholder investment, emphasizing instead community self-determination and the mutualization of benefits.[47]

Such projects, Bollier argues, are part of a transition from homo economicus to commoner, or from consumer to citizen. This would realign the WGW with the original ambitions of the World Game. David McConville, Chair of the Board of the Buckminster Fuller Institute, notes the World Game was one of Fuller's 'more profoundly subversive visions' because its status as a game made it a widely accessible tool. Its ambition was to be widely and freely disseminated. 'The good ol' boy political process was to be subverted out of existence by a process that brought distributed scenario gameplay and decision making to the twentieth century'.[48] A further question is whether such a subversive vision can or should adapt to twenty-first century demands for 'Global System Change' and the transition to a 'Steady State Economy'.[49]

Notes

1 Medard Gabel, 'Global Recall and Networld Game: Global Access to Global Models', in *Integrated Global Models of Sustainable Development - Volume III*, ed. Akira Onishi (Oxford: EOLSS Publications, 2009), unpaginated. Available online at http://www.eolss.net/sample-Chapters/C15/E1-47-18.pdf.

2 Gabel, 'Global Recall and Networld Game', unpaginated.

3 Gert Biesta, *Letting Art Teach: Art Education After Beuys* (Arnhem: Artez Press, 2017).

4 Jon Dieges and Edwin Schlossberg, Good News: Journal of Environmental Design, 1968, unpaginated, Stanford University Collections M1090, Series 18: Project Files, World Game Subseries 2, box 105, Folder 2: Tom Turner Files, World Game Class.

5 Daryl E. Lembke, 'Mayor of Berkeley Spat Upon: Berkeley Put under Curfew after 3rd Day of Violence Tight Curfew for Berkeley Put in Effect', *Los Angeles Times*, 1 July 1968, 1.

6 Dieges and Schlossberg, Good News: Journal of Environmental Design.

7 See Timothy Stott, 'Ludic Pedagogies at the College of Environmental Design, UC Berkeley, 1966 to 1972', in *The Culture of Nature in the History of Design*, ed. Kjetil Fallan (London: Routledge, 2019).

8 Harvard W. McLean, 'Simulation Games: Tools for Environmental Education', *The Elementary School Journal* 73, no. 7 (April 1973), 375.

9 John Hunter and Jamie Field Baker, 'Teaching for a Safer World', *Independent School* 72, no. 7 (Winter 2013), 44.

10 Hunter and Baker, 'Teaching for a Safer World', 45.

11 John Hunter, letter to parents and teachers, Richmond Public Schools, undated (late summer 1978).

12 Hunter, letter to parents and teachers.

13 Warren Bennis and Burt Nanus, *Leaders: The Strategies for Taking Charge* (New York: Harper & Row, 1985).

14 John Hunter, *World Peace Game and Other 4th-Grade Achievements* (Boston, MA: Houghton Mifflin Harcourt, 2013), 8.

15 Hunter in Krystal Goree, 'An Interview with John Hunter: Unleashing Potential through Open Space Thinking', *Tempo* XXXVIII, no. 3 (2017), 6.

16 Hunter, *World Peace Game*, 18.

17 Hunter, *World Peace Game*, 24.

18 Hunter and Baker, 'Teaching for a Safer World', 45.

19 Hunter and Baker, 'Teaching for a Safer World', 44.

20 Hunter, *World Peace Game*, 250–251.

21 Paul-Michael Dekker, Peter Meisen, and Amy B. Bruton, 'The GENI Model: The Interconnection of Global Power Resources to Obtain an Optimal Global Sustainable Energy Solution', *Simulation* 64 (April 1995), 253.

22 R. Buckminster Fuller, *Critical Path* (New York: St. Martin's Press, 1981), 199.

23 For a full GENI history, see http://www.geni.org/globalenergy/issues/overview/history.shtml.

24 Dekker et al., 'The GENI Model', 252.

25 GENI, Global Model Index, http://www.geni.org/globalenergy/library/geni/globalmodelindex.shtml.

26 Mark Wasiuta, 'The Persistence of Informational Vision: World Game 1969, 2009', *Journal of the Society of Architectural Historians* 68, no. 4 (2009), 593.

27 The World Game Corporate Program, *Facilitator/Guide Script* (Philadelphia: World Game Institute, 1993), 3.

28 World Game Workshop Teacher's Manual (Philadelphia: World Game Institute, 1994), 7.

29 World Game Workshop Teacher's Manual, unpaginated.

30 The World Game Corporate Program, 24.

31 World Game Workshop Teacher's Manual, 12.

32 The World Game Corporate Program, 55.

33 o.s.Earth Inc, LinkedIn profile page: https://www.linkedin.com/company/o.s.earth-inc./about/.

34 BigPicture Consulting, 'The FutureGame Simulation', http://www.bigpictureconsulting.com/future_game/.

35 Medard Gabel and Jim Walker, 'The Anticipatory Leader: Buckminster Fuller's Principles for Making the World Work', *The Futurist* 40, no. 5 (September–October 2006), 40.

36 George Melloan, 'Feeling the Muscles of the Multinationals', *Wall Street Journal*, 6 January 2004, A19.

37 Christophe Bonneuil and Jean-Baptiste Fressoz, *The Shock of the Anthropocene: The Earth, History and Us*, trans. David Fernbach (London: Verso, 2016), 225.

38 https://centerforneweconomics.org/apply/civic-synergy-program/

39 R. Buckminster Fuller, *Ideas and Integrities: A Spontaneous Autobiographical Disclosure* (New York: Macmillan, 1963), 272.

40 Greg Watson, 'Education for a Design Science Revolution: Buckminster Fuller's World Game Workshop', unpublished presentation at Rhode Island School of Design, 5 November 2020.

41 Juliette N. Rooney-Varga, Florian Kapmeier, John D. Sterman, Andrew P. Jones, Michele Putko, and Kenneth Rath, 'The Climate Action Simulation', *Simulation & Gaming* 51, no. 2 (April 2020), 114–140. doi:10.1177/1046878119890643.

42 Jennifer Gabrys, *Program Earth: Environmental Sensing Technology and the Making of a Computational Planet* (Minneapolis: University of Minnesota Press, 2016), 4, 268.

43 Gabrys, *Program Earth*, 274.

44 See Vasilis Kostakis, Vasilis Niaros, George Dafermos, and Michel Bauwens, 'Design Global, Manufacture Local: Exploring the Contours of an Emerging Productive Model', *Futures* 73 (2015), 126–135.

45 The previous two paragraphs draw extensively from Greg Watson in conversation with the author, Monday 18 January 2021.

46 David Bollier, 'The Next Big Thing Will Be a Lot of Small Things', *The Nation* (28 August/4 September, 2017), 18.

47 Bollier, 'The Next Big Thing', 17.

48 David McConville, 'The Consequences of Various World Plans …', *Ubiquity: The Journal of Pervasive Media* 1, no. 1 (2012), 79. doi:10.1386/ubiq.1.1.65_1

49 Ann Pettifor, *The Case for a Green New Deal* (London: Verso, 2019).

Conclusion

In conclusion, we might look at world gaming as a sub-category of so-called gamification, or the application of ludic elements to non-game contexts to encourage user engagement. When first coined in 2008, the term described how companies turn their products and services into entertainment platforms, but is now widespread across marketing, voting systems, education, exercise and health, UX design and participatory design, and even political activity.[1] Jane McGonigal, for example, a game designer from California who has worked as a consultant for MacDonald's and the Olympic Games Committee, believes gamification, in addition to generating profit, can be a problem-solving tool applied by gamers to a range of social, ecological, and political problems.[2] Advocates of world gaming share this view.

Acting politically through gaming, however, should give us pause. As many of the examples discussed in this book show, there is a close similarity between a 'choice architecture', that is, an architecture used in behavioural economics in which subjects are guided to make profitable decisions, and the structure of a game. In both cases, 'participants can only choose between the options given to them'. The aim is not to 'change mindsets but only visible and measurable performance and conduct'.[3] This returns us to the question of how and to what extent reformation of the game environment might correlate with the 'political expedient of attempting to reform man' abandoned by Fuller. Game worlds are, operationally, at least, closed, as shown by the recent consolidation of filter bubbles as game worlds and the subsequent rise of the likes of QAnon, the latter having many of the features of an 'alternate-reality game', or ARG, defined to be a 'clue-cracking, multiplatform scavenger hunt'.[4] In light of this, the promise that a 'ludic century' will follow the information revolution of the twentieth century need not give cause for celebration.[5]

Conversely, however, gamification might promote participatory sensing and citizen science. A recent study shows how gamification might 'enhance human engagement and user experience in participatory sensing systems for environmental monitoring' and, in doing so, close the gap between 'researchers, environmental experts, decision-makers, and the people, while collecting data and building a whole new level of services (from the people, for the people)'.[6] Such studies return us to the present potential of world gaming and the repurposing of data for egalitarian, educational, and ecological purposes. This book has traced the evolution of the World Game and other world gaming experiments simply to show both the limitations and necessity of designing a political arena for planetary management, an arena that can be pedagogical and might be operational, but which still must grapple with design as a key determinant of how problems are conceived, who can address them, and how. Such an arena would need to be multi-scalar, distributed, accessible, and pluriversal, if it is to face the wickedness – even 'super wickedness' – of problems of environmental design, planning, and policy under current conditions.[7]

Notes

1 Sebastian Deterding, Rilla Khaled, Lennart Nacke, and Dan Dixon, 'Gamification: Toward a Definition', *CHI '11: Proceedings of the 2011 SIGCHI Conference on Human Factors in Computing Systems*, May 2011, https://dl.acm.org/doi/proceedings/10.1145/1978942.

2 Jane McGonigal, *Reality Is Broken: Why Games Make Us Better and How They Can Change the World* (New York: Penguin, 2011).

3 Niklas Schrape, 'Gamification and Governmentality', in *Rethinking Gamification*, eds. S. Fizek, M. Fuchs, P. Ruffino, and N. Schrape (Lüneburg: Meson Press, 2014), 35.

4 Clive Thompson, 'QAnon Is like a Game—A Most Dangerous Game', *Wired*, 22 September 2020, https://www.wired.com/story/qanon-most-dangerous-multiplatform-game/.

5 Eric Zimmerman, 'Position Statement: Manifesto for a Ludic Century', in *The Gameful World: Approaches, Issues, Applications*, eds. Walz Steffen P. and Deterding Sebastian (Cambridge: MIT Press, 2014), 19–22.

6 Maria V. Palacin-Silva, Antti Knutas, Maria Angela Ferrario, Jari Porras, Jouni Ikonen, and Chandara Chea, 'The Role of Gamification in Participatory Environmental Sensing: A Study in the Wild', *Proceedings of the 2018 CHI Conference on Human Factors in Computing Systems (CHI '18)*, 2018, Association for Computing Machinery, New York, Paper 221, 9, 10.

7 See Kelly Levin, Benjamin Cashore, Steven Bernstein, and Graeme Auld, 'Overcoming the Tragedy of Super Wicked Problems: Constraining Our Future Selves to Ameliorate Global Change', *Policy Science* 45 (2012), 123–152.

Bibliography

Abram, Tom. Interview with Chris Winter, 9 January 2020. https://archivesit.org.uk/interviews/chris-winter/

Adair, Wendy. 'International Year of the Child Hailed by New UHCC Global Report'. *UH Central*, 5 January 1979.

Anker, Peder. *From Bauhaus to Ecohouse: A History of Ecological Design*. Baton Rouge: Louisiana State University Press, 2010.

Anon. 'Cohen 'Ecstatic' Over Attracting Major Center'. *Reporter* 11, no. 3, 20 September 1979: 2.

Anon. 'Handwritten notes on World Game Package', Stanford University Collections M1090, Series 18: Project Files, World Game Subseries 2, box 39, Folder 7, undated and unpaginated.

Anon. Press release. *Colonial News*, Thursday 14 April 1970: 2.

Anon. 'R. Buckminster Fuller's Dymaxion World'. *Life*, 1 March 1943: 41–55.

Anon. 'Research in the Present about the Future'. *Colonial News*, Tuesday 5 March 1970: 5.

Anon. 'SUAB Director Authors Book on Ecology'. *Vestal News*, 24 November 1970.

Anon. 'SUAB Professor on TV Show about the Future'. *Press* (Binghamton, NY), Wednesday 5 February 1969.

Anon. 'SUNY Gets $20,000 UN Grant'. *The Sun Bulletin*, 19 February 1974.

Anon. 'The Woods Society'. *Pipe Dream*, Friday 30 November 1973: 11.

Anonymous. 'World Game Pattern Recognition System for Resource Utilisation Planning'. Undated. Stanford University Collections M1090, Series 18: Project Files, World Game Subseries 2, box 27, Folder 2: World Game History, General (2 of 2), 2–3.

Ashby, W. Ross. *An Introduction to Cybernetics*. London: Chapman & Hall, 1956.

Ashley, Richard. 'The Eye of Power: The Politics of World Modelling'. *International Organization* 37, no. 3 (1983): 495–535.

Asker, Jim. 'UH Think Tank Involved in Study of World's Ills'. *The Houston Post*, Sunday 5 June 1977, 8B.

Bateson, Gregory. 'The Message —This Is Play'. *Transactions of the Second Conference on Group Processes* 2 (October 1955): 145–242.

Bateson, Gregory. 'A Theory of Play and Fantasy'. In *Steps to an Ecology of Mind: Collected Essays in Anthropology, Psychiatry, Evolution, and Epistemology*. Chicago, IL: University of Chicago Press, 1972, 177–193.

Bender, John and Michael Marriman. *The Culture of Diagram*. Stanford, CA: Stanford University Press, 2010.

Ben-Eli, Michael and Edwin Schlossberg. Brochure for The World Game for Government Executives. 14–18 December 1970 and 1–5 March 1971. Unpaginated. Stanford University Collections M1090, Series 18: Project Files, World Game Subseries 2, box 65, Folder 4: World Game Correspondence, Locations – Washington, DC.

Bennis, Warren and Burt Nanus, *Leaders: The Strategies for Taking Charge*. New York: Harper & Row, 1985.

Bernard, Thérèse, ed. *Expo 67, Official Guide*. Toronto: Maclean-Hunter Publishing, 1967.

Berne, Eric. *Games People Play: The Psychology of Human Relationships*. New York: Grove Press, 1964.

Bessette, Juliette. 'John McHale: de l'art du collage à la pensée prospective'. *Les Cahiers du Mnam* 140 (2017): 34–58.

Biermann, Frank. *Earth System Governance: World Politics in the Anthropocene*. Cambridge, MA: The MIT Press, 2014.

Biesta, Gert. *Letting Art Teach: Art Education after Beuys*. Arnhem: Artez Press, 2017.

BigPicture Consulting, 'The FutureGame Simulation'. http://www.bigpictureconsulting.com/future_game/

Blyth, Tilly. 'Narratives in the History of Computing: Constructing the Information Age Gallery at the Science Museum'. In *Making the History of Computing Relevant*, edited by A. Tatnall, T. Blyth and R. Johnson. HC 2013. IFIP Advances in Information and Communication Technology 416 (2013), 25–34. doi:10.1007/978-3-642-41650-7.

Bollier, David. 'The Next Big Thing Will Be a Lot of Small Things'. *The Nation* (28 August/4 September 2017): 16–25.

Bonneuil, Christophe and Jean-Baptiste Fressoz. *The Shock of the Anthropocene: The Earth, History and Us*. Translated by David Fernbach, London: Verso, 2016, 33.

Boulding, Kenneth E. 'Earth as a Spaceship'. Address to NASA Committee on Space Sciences, Washington State University, 10 May 1965. Box 38, Kenneth E. Boulding Papers, University of Colorado at Boulder Libraries.

Boulding, Kenneth E. 'The Economics of the Coming Spaceship Earth'. In *Environmental Quality in a Growing Economy*, edited by H. Jarrett. Baltimore, MD: Johns Hopkins University Press, 1966, 3–14.

Bourdon, Pierre. Letter to John McHale, 5 June 1967. Stanford University Collections M1090, Series 18: Project Files, World Game Subseries

2, box 62, Folder 5: World Game Correspondence, Locations – McGill University (Expo 67).

Brand, Stewart. 'It Began with World War IV'. In *The New Games Book*, edited by A. Fluegelman. New York: Headlands Press, 1976, 7–20.

Brown, Howard. 'New Business Models, Measurement, and Methodologies'. In *Evolutions in Sustainable Investing: Strategies, Funds, and Thought Leadership*, edited by C. Krosinsky, N. Robins and S. Viederman. Hoboken, NJ: John Wiley & Sons, 2012, 423–428.

Bruton, Amy B., Paul-Michael Dekker and Peter Meisen, 'The GENI Model: The Interconnection of Global Power Resources to Obtain an Optimal Global Sustainable Energy Solution'. *Simulation* 64 (April 1995): 244–252.

Clarke, Bruce and Mark B. N. Hansen, eds. *Emergence and Embodiment: New Essays on Second-Order Systems Theory*. Durham, NC: Duke University Press, 2009.

Clarke, Bruce. 'Rethinking Gaia: Stengers, Latour, Margulis'. *Theory, Culture & Society* 34, no. 4 (2017): 3–26.

Colomina, Beatriz. 'Enclosed by Images: The Eameses' Multimedia Architecture'. *Grey Room* 2 (Winter 2001): 5–29.

Coulter, Bill. 'Millions of Children Face Bleak Future, UH Study Finds'. *The Houston Post*, Sunday 11 March 1979, 3D.

Christakis, Alexander. Review in *Technological Forecasting* 1 (1970): 427.

Csikszentmihalyi, Mihaly and Stith Bennett. 'An Exploratory Model of Play'. *American Anthropologist* 73, no. 1 (1971): 45–58.

Dalmedico, Amy and Matthias Heymann. 'Epistemology and Politics in Earth System Modelling: Historical Perspectives'. *Journal of Advances in Modelling Earth Systems* 11 (2019): 1139–1152.

Davis, Gerald and Ralph Rinzler. *Programme for 1972 Festival of American Folklife*. New York: Music Sales Corporation, 1972.

Davis, Robert H. 'International Influence Process: How Relevant Is the Contribution of Psychologists?' In *Psychological Research in National Defense Today*, edited by J. E. Ohlaner. Technical Report S-1, US Army Behavioural Science Research Laboratory, June 1967.

Deese, R. S. 'The Artifact of Nature: 'Spaceship Earth' and the Dawn of Global Environmentalism'. *Endeavour* 33, no. 2 (June 2009): 70–75.

Deterding, Sebastian, Rilla Khaled, Lennart Nacke and Dan Dixon. 'Gamification: Toward a Definition'. *CHI '11: Proceedings of the SIG-CHI Conference on Human Factors in Computing Systems*, May 2011, doi:10.1145/1978942

Dieges, Jon and Edwin Schlossberg. *Good News: Journal of Environmental Design*, 1968. Stanford University Collections M1090, Series 18: Project Files, World Game Subseries 2, box 105, Folder 2: Tom Turner Files, World Game Class).

Donovan, Patricia, 'Magda Cordell McHale, Professor Emerita'. 26 February 2008. http://www.buffalo.edu/news/releases/2008/02/9185.html

Drucker, Johanna. *Graphesis: Visual Forms of Knowledge Production.* Cambridge: Harvard University Press, 2014.

Edwards, Paul N. *A Vast Machine: Computer Models, Climate Data and the Politics of Global Warming.* Cambridge, MA: MIT Press, 2010.

Escobar, Arturo. *Designs for the Pluriverse: Radical Interdependence, Autonomy, and the Making of Worlds.* Durham, NC: Duke University Press, 2018.

Farrell, Barry. 'The View from the Year 2000'. *Life*, 26 February 1971: 50–53.

Fernandez, Maria. "Aesthetically Potent Environments', or How Gordon Pask Detourned Instrumental Cybernetics'. In *White Heat Cold Logic: British Computer Art 1960–1980*, edited by Paul Brown, Charlie Gere, Nicholas Lambert and Catherine Mason. Cambridge, MA: The MIT Press, 2009, 53–70.

Ferran, Bronac and Elisabeth Fisher. 'The Experimental Generation: Networks of Interdisciplinary Praxis in Post War British Art (1950–1970)'. *Interdisciplinary Science Reviews* 42, nos. 1–2 (2017): 1–3.

Forrester, Jay. *Industrial Dynamics.* Waltham, MA: Pegasus, 1961.

Forrester, Jay. 'Counterintuitive Behavior of Social Systems'. *Simulation* 16, no. 2 (1971): 61–76.

Forrester, Jay. *World Dynamics*, Second Edition. Cambridge, MA: Wright-Allen Press, 1973.

Forrester, Jay. 'The Beginning of System Dynamics'. Banquet Talk at the International Meeting of the System Dynamics Society, Stuttgart, 13 July 1989. https://web.mit.edu/sysdyn/sd-intro/D-4165-1.pdf

Foundation General Systems Ltd. 'A Description of the Computer 70 Theme Exhibit, Olympia, London 5–9 October 1970'. *PAGE* 16, June 1971, unpaginated.

Fry, Tony. *Design as Politics.* Oxford: Berg, 2010.

Fulford, Robert. *Remember Expo: A Pictorial Record.* Toronto: McClelland and Stewart, 1968.

Fuller, R. Buckminster. *50 Years of the Design Science Revolution and the World Game.* Carbondale: World Resources Inventory, Southern Illinois University, 1969.

Fuller, R. Buckminster cartography, US patent 23939676, filed 25 February 1944, issued 29 January 1946.

Fuller, R. Buckminster. *Critical Path.* New York: St. Martin's Press, 1981.

Fuller, R. Buckminster. Elevation for the World Resources Simulation Center, Phase I. c. 1969. Stanford University Collections M1090, Series 18: Project Files, World Game Subseries 2, 32, Folder 2: World Game History II, World Game Report (1 of 2).

Fuller, R. Buckminster. Hearings Before the Committee on Foreign Relations, United States Senate, Ninety-Fourth Congress, First Session of the United States and the United Nations and the Nomination of Daniel Patrick Moynihan to be US Representative to the United Nations with the Rank of Ambassador. Washington, DC: US Government Printing Office, 1975.

Fuller, R. Buckminster. *Ideas and Integrities: A Spontaneous Autobiographical Disclosure*. New York: Macmillan, 1963.

Fuller, R. Buckminster. Keynote address to How To Make The World Work, conference at SIU in October 1965, Stanford University Collections M1090, Series 18: Project Files, World Game Subseries 2, box 39, Folder 11: Vision 65 Address, 8.

Fuller, R. Buckminster. *No More Secondhand God*. New York: Anchor Books, 1971.

Fuller, R. Buckminster. *Operating Manual for Spaceship Earth*. Edwardsville: Southern Illinois University Press, 1968.

Fuller, R. Buckminster. 'Planned Implementation of the World Resources Simulation Center, Edwardsville, Illinois'. Presentation to the Joint National Meeting of the American Aeronautical Society and Operations Research Society, Brown Palace and Cosmopolitan Hotels, Denver, CO. 17–20 June 1969. Paper given on Wednesday 18 June, 12pm to 2.30pm. Stanford University Collections M1090, Series 18: Project Files, World Game Subseries 2, box 25, Folder 1: World Game History, World Resources Simulation Center (2 of 3).

Fuller, R. Buckminster. 'The World Game'. Presentation to the Joint National Meeting of the American Aeronautical Society and Operations Research Society, Brown Palace and Cosmopolitan Hotels, Denver, CO. 17–20 June 1969. Paper given on Wednesday 18 June, 12pm to 2.30pm. Stanford University Collections M1090, Series 18: Project Files, World Game Subseries 2, box 25, folder 11.

Fuller, R. Buckminster. Typescript for 'World Game: How It Came About', 21 April 1968. Stanford University Collections M1090, Series 18: Project Files, World Game Subseries 2, box 27, Folder 2: World Game History, General (2 of 2), 5.

Fuller, R. Buckminster. Undated presentation at SIU. Undated. Stanford University Collections M1090, Series 18: Project Files, World Game Subseries 2, box 24, Folder 14: World Game History, World Game Presentation SIU.

Fuller, R. Buckminster. World Resources Simulation Center, Stanford University Collections M1090, Series 18: Project Files, World Game Subseries 2, box 25, Folder 1.

Gabel, Medard. 'Buckminster Fuller's World Game'. *Whole Earth Catalog*, March 1970: 31.

Gabel, Medard. 'World Game 'World View'/Frames of Reference Are Composed Of...', 5 October 1970. Stanford University Collections M1090, Series 18: Project Files, World Game Subseries 2, box 27, Folder 2, 3.

Gabel, Medard. 'Global Recall and Networld Game: Global Access to Global Models'. In *Integrated Global Models of Sustainable*

Development - Volume III, edited by Akira Onishi. Oxford: EOLSS Publications, 2009. http://www.eolss.net/sample-Chapters/C15/E1-47-18.pdf

Gabel, Medard and Jim Walker. 'The Anticipatory Leader: Buckminster Fuller's Principles for Making the World Work'. *The Futurist* 40, no. 5 (September-October 2006): 39–44.

Gabrys, Jennifer. *Program Earth: Environmental Sensing Technology and the Making of a Computational Planet*. Minneapolis: University of Minnesota Press, 2016.

Galison, Peter. 'The Ontology of the Enemy: Norbert Wiener and the Cybernetic Vision'. *Critical Inquiry* 21, no. 1 (Autumn 1994): 228–266.

Goree, Krystal. 'An Interview with John Hunter: Unleashing Potential through Open Space Thinking'. *Tempo* XXXVIII, no. 3 (2017): 1–11.

Hansen, Mark Victor. 'An Aerial View of Ecology by World Game'. *Student Handbook for International Design Conference*, Aspen 1970. Stanford University Collections M1090, Series 18: Project Files, World Game Subseries 2, box 14, 32/3: World Game History II, World Game Report (2 of 2).

Heaviwait, Rebus and Emmanuel Lighthanger. *Projex*. New York: Links Books, 1972.

Highmore, Ben. 'Machinic Magic: IBM at the 1964–1965 New York World's Fair'. *New Formations* 51 (Winter 2003): 128–148.

Highmore, Ben. *The Art of Brutalism: Rescuing Hope from Catastrophe in 1950s Britain*. New Haven, CT: Yale University Press, 2017.

Höhler, Sabine. '"Spaceship Earth": Envisioning Human Habitats in the Environmental Age'. *GHI Bulletin* 42 (Spring 2008): 65–85.

Holland, Owen and Phil Husbands. 'The Origins of British Cybernetics: The Ratio Club'. *Kybernetes* 40, nos 1/2 (2011): 110–123.

Hunter, John. Letter to parents and teachers, Richmond Public Schools. Undated (late summer 1978).

Hunter, John. *World Peace Game and Other 4th-Grade Achievements*. Boston, MA: Houghton Mifflin Harcourt, 2013.

Hunter, John and Jamie Field Baker. 'Teaching for a Safer World'. *Independent School* 72, no. 7 (Winter 2013): 42–46.

Huizinga, Johann. *Homo Ludens: A Study of the Play Element in Culture*. Boston, MA: Beacon Press, 1955.

Ingold, Tim. *The Perception of the Environment*. London: Routledge, 2000.

King, Alexander and Bertrand Schneider. *The First Global Revolution*. New York: Pantheon, 1991.

Kirk, Andrew. *Counterculture Green: The Whole Earth Catalog and American Environmentalism*. Lawrence: University Press of Kansas, 2007.

Kitnick, Alex, ed. *John McHale – The Expendable Reader: Articles on Art, Architecture, Design, and Media*. New York: GSAPP Books, 2011.

Kopstein F. F. and Isabel J. Shillestad. 'A Survey of Auto-Instructional Devices'. *Aeronautical Services Division Technical Report* 61–414, September 1961, AD 268223.

Kostakis, Vasilis, Vasilis Niaros, George Dafermos and Michel Bauwens. 'Design Global, Manufacture Local: Exploring the Contours of an Emerging Productive Model'. *Futures* 73 (2015): 126–135.

Kouw, Matthijs, Christoph van den Heuvel and Andrea Scharnhorst. 'Exploring Uncertainty in Knowledge Representations: Classifications, Simulations, and Models of the World'. In *Virtual Knowledge. Experimenting in the Humanities and Social Sciences*, edited by P. Wouters, A. Beaulieu, A. Scharnhorst and S. Wyatt. Cambridge, MA: The MIT Press, 2012.

La Tourette, John E. Letter to John McHale, 10 November 1975. Special Collections, Binghamton University, State University of New York.

Landry, Bebe. *SUNY Binghamton News*, 9 September 1971: 30.

Landry, Bebe. *SUNY Binghamton News*, 5 January 1972: 30.

Landry, Bebe. *SUNY Binghamton News*, 4 September 1973: 30.

Landry, Bebe. *SUNY Binghamton News*, 11 September 1974: 30.

Lansdown, John. 'The Name of the Game Is...? A Personal View of the Computer Arts Society's Project'. *The Computer Bulletin* 14, no. 9 (September 1970), unpaginated.

Lansdown, John. 'Computer Graphics ≠ Computer Art'. *PAGE* 19 (December 1971): 2.

Latour, Bruno. *Re-Assembling the Social: An Introduction to Actor-Network-Theory*. Oxford: Oxford University Press, 2005.

Latour, Bruno. 'Why Gaia Is Not a God of Totality'. *Theory, Culture & Society* 34, nos. 2–3 (2017): 61–81.

Lawyer Phillip H. and Richard Meyer. Letter to Dr Marty Grober, 19 August 1970. Stanford University Collections M1090, Series 18: Project Files, World Game Subseries 2, box 65, Folder 4: World Game Correspondence, Locations – SIU, Free School.

Lee, Pamela. *New Games: Postmodernism after Contemporary Art*. London: Routledge, 2013.

Lembke, Daryl E. 'Mayor of Berkeley Spat Upon: Berkeley Put Under Curfew after 3rd Day of Violence Tight Curfew for Berkeley Put in Effect'. *Los Angeles Times*, 1 July 1968: 1.

Letter from Ackerman to Tom Turner, 23 April 1970, Stanford University Collections M1090, Series 18: Project Files, World Game Subseries 2, box 62, Folder 6: World Game Correspondence, Locations, Montreal (1 of 2).

Levin, Kelly, Benjamin Cashore, Steven Bernstein and Graeme Auld. 'Overcoming the Tragedy of Super Wicked Problems: Constraining Our Future Selves to Ameliorate Global Change'. *Policy Science* 45 (2012): 123–152.

Linstone, Harold A. 'Editorial Comment: The Rome Special World Conference on Futures Research'. *Technological Forecasting and Social Change* 5, no. 4 (1973): 329.

Lövbrand, Eva, Johannes Stipple and Bo Wiman. 'Earth System Governmentality: Reflections on Science in the Anthropocene'. *Global Environmental Change* 19, no. 1 (2008): 7–13.

Lovelock, James. *Gaia: The Practical Science of Planetary Medicine*. Oxford: Oxford University Press, 1991.

Lowe, Robert. 'Future Tense or Future Perfect?' *Pipe Dream*, Friday 22 November 1974: 21.

MacGonigal, Jane. *Reality Is Broken: Why Games Make Us Better and How They Can Change the World*. New York: Penguin, 2011.

Mackinder, Harold. 'The Geographical Pivot of History'. *The Geographical Journal* XXIII, no. 4 (April 1904): 421–437.

Mackinder, Harold. *Democratic Ideals and Reality: A Study in the Politics of Reconstruction*. London: Constable & Co, 1919.

Mallen, George. 'Bridging Computing in the Arts and Software Department'. In *White Heat Cold Logic: British Computer Art 1960–1980*, edited by Paul Brown, Charlie Gere, Nicholas Lambert and Catherine Mason. Cambridge, MA: The MIT Press, 2008, 191–202.

Mallen, George /Computer Arts Society. 'Ecogame', September 1970. Stanford University Collections M1090, Series 18: Project Files, World Game Subseries 2, box 62, Folder 2: World Game Correspondence, Locations, London.

Mallen, George. 'Early Computer Models of Cognitive Systems and the Beginnings of Cognitive Systems Dynamics'. *Constructivist Foundations* 9, no. 1 (2013): 137–138.

Mallen, George. 'Ecogame: Computing in a Cultural Context'. unpublished draft paper, September 2019.

Marks, Spencer. 'Co-Op City Community Study', 1970. Stanford University Collections M1090, Series 18: Project Files, World Game Subseries 2, box 39, Folder 16: World Game History, Val Winsey (1 of 3).

Martin, Rheinhold. *The Organisational Complex: Architecture, Media, and Corporate Space*. Cambridge, MA: MIT Press, 2003.

Martin, Reinhold. 'Fuller's Futures'. In *New Views on R. Buckminster Fuller*, edited by Hsiao-Yun Chu and Roberto G. Trujillo. Stanford, CA: Stanford University Press, 2009, 176–187.

Maslow, A. H. 'A Theory of Metamotivation: The Biological Rooting of the Value-Life'. *Journal of Humanistic Psychology* 7, no. 2 (1967): 93–127.

Mason, Catherine. 'The Fortieth Anniversary of Event One at the Royal College of Art'. Paper delivered at Electronic Visualisation and the Arts annual conference, 6 to 8 July 2009. Published online via ScienceOpen by the BCS. https://www.scienceopen.com/hosted-document?doi=10.14236/ewic/EVA2009.15. doi:10.14236/ewic/EVA2009.15

Massey, Jonathan. 'Buckminster Fuller's Cybernetic Pastoral: The United States Pavilion at Expo 67'. *The Journal of Architecture* 11, no. 4 (2006): 463–483.

Matter, Herbert. *The World Game: The Structure of Nature* (I'm going to take you to breakfast yesterday morning). Filmed at 1969 NY Studio School, World Game Seminar, Saturn Pictures, 23–30 June 1969. 60 min., b & w, 16 mm. Stanford University Collections M1090, Series 17: Subseries 7, 73a.

Matter, Herbert. *The World Game: Playing World Game (A Hundred Million Horses Going Nowhere)*. Filmed at 1969 NY Studio School, World Game Seminar, Saturn Pictures, 23–30 June 1969. 60 min., b & w, 16 mm. Stanford University Collections M1090, Series 17: Subseries 7, 73g.

Matter, Herbert. *The World Game: Playing World Game (Getting Power to the People)*. Filmed at 1969 NY Studio School, World Game Seminar, Saturn Pictures, 23–30 June 1969. 60 min., b & w, 16 mm. Stanford University Collections M1090, Series 17: Subseries 7, 73i.

Matter, Herbert. *World Game Can Work (Politicians Will Yield to the Computer)*. Filmed at 1969 NY Studio School, World Game Seminar, Saturn Pictures, 23–30 June 1969. 60 min., b & w, 16 mm. Stanford University Collections M1090, Series 17: Subseries 7, 73j.

McCall, Anthony. *PAGE* 16 (June 1971): unpaginated.

McConville, David. 'The Consequences of Various World Plans …'. *Ubiquity: The Journal of Pervasive Media* 1, no. 1 (2012): 65–80.

McFarland, Gay E. 'The Future: Couple Share a Job of Peering at Years Ahead'. *Houston Chronicle*, Sunday 11 June 1978: 20.

McHale, John. Statement. Private view card for exhibition '3 Collagists - new work by E L T Mesens, John McHale, Gwyther Irwin' (recto). Institute of Contemporary Arts, London, 5–29 November 1958.

McHale, John. Untitled statement in brochure Paris 1965, World Design Science Decade. Stanford University Collections M1090, Series 18: Project Files, World Game Subseries 2, box 62, Folder 8: World Game Correspondence, Locations, Paris (1965).

McHale, John. 'General Description of the Sketch Model', February 1966. Stanford University Collections M1090, Series 18: Project Files, World Game Subseries 2, box 105, Folder 9: Tom Turner Files, World Game Facility.

McHale, John. *Phase II (1967), Document 6, The Ecological Context: Energy and Materials*. Carbondale, IL: World Resources Inventory, 1967.

McHale, John. *The Ecological Context*. New York: George Braziller, 1970.

McHale, John. 'The Changing Patterns of Futures Research in the USA'. *Futures* 5, no. 3 (June 1973): 257–271.

McHale, Magda Cordell. Reflections on Impacts of the Information Environment. Prepared for the Intergovernmental Bureau for Informatics, Rome, 1 June 1982. University of Buffalo Special Collections, Center For Integrative Studies 1982–83, Folder 1.

McHale, Magda Cordell. Summary Conference Report of the International Conference on Information Revolution: Different Socio-Cultural Perceptions, April 1983, University of Buffalo Special Collections, Center For Integrative Studies 1982–83, Folder 1.

McHale, Magda Cordell and John McHale. 'World Trends and Alternative Futures'. *Open Grants Papers* no. 1 (Honolulu: East-West Center, 1974).

McHale, Magda Cordell and John McHale. 'Future Studies: An International Survey'. *Ekistics* 41, no. 246 (May 1976): 300–306.

McHale, Magda Cordell and John McHale. *Children in the World.* Washington, DC: Population Reference Bureau, 1979.

McHale, Magda Cordell and John McHale, with Guy Streatfield and Laurence Tobias, eds. *The Futures Directory: An International Listing and Description of Organizations and Individuals Active in Futures Studies and Long-Range Planning.* Guildford: IPC Science and Technology Press, 1977.

McHale, John, Magda Cordell McHale and Guy Streatfield. *Women in World Terms: Facts and Trends.* Houston, TX: University of Houston, Center for Integrative Studies, 1975.

McKinnon-Wood, Robin. 'Early Machinations'. *Systems Research* 10, no. 3 (1993): 129–132.

McLean, Harvard W. 'Simulation Games: Tools for Environmental Education'. *The Elementary School Journal* 73, no. 7 (April 1973): 374–380.

Meadows, Dennis, Donella Meadows and Jørgen Randers. *Beyond the Limits: Global Collapse or a Sustainable Future.* London: Earthscan, 1992.

Meadows, Dennis, Donnella Meadows, Jørgen Randers and William W. Behrens III. *The Limits to Growth: A Report on the Club of Rome's Project on the Predicament of Mankind.* New York: Universe Books, 1972.

Medina, Eden. *Cybernetic Revolutionaries: Technology and Politics in Allende's Chile.* Cambridge, MA: The MIT Press, 2011.

Melloan, George. 'Feeling the Muscles of the Multinationals'. *Wall Street Journal,* 6 January 2004: A19.

Miller, Bob. Letter to John McHale, 14 October 1971. Binghamton University Archives, Special Collections, Binghamton University Libraries, Binghamton University, State University of New York.

Moss, Lawrence A. 'Sat: A Small Glimpse of the Future Today'. *The Colonial News,* Friday 13 March 1970: 4.

Nake, Frieder. 'Technocratic Dadaists'. *PAGE* 21 (March 1972): unpaginated.

Nake, Frieder. 'The Semiotic Engine: Notes on the History of Algorithmic Images in Europe'. *Art Journal* 68, no. 1 (Spring 2009): 76–89.

Nelson, Carl G. *Progress Report on Cooperative Research in World Design.* Stanford University Collections M1090, Series 18: Project Files, World Game Subseries 2, box 24, Folder 15: Progress Report (Carl G. Nelson), 1 April 1969.

Nelson, Robert Colby. 'Nature's Extraordinary Order'. *Christian Science Monitor,* Tuesday 3 November 1964.

Nieland, Justus. 'Midcentury Futurisms: Expanded Cinema, Design, and the Modernist Sensorium'. *Affirmations: Of the Modern* 2, no. 1 (2014): 46–84.

Nisbet, James. *Ecologies, Environments and Energy Systems in Art of the 1960s and 1970s.* Cambridge, MA: MIT Press, 2014.

Office of the President (SIU). 'Notes Concerning Plans for the Establishment of the World Resources Computing Center', 1 May 1968. Stanford University Collections M1090, Series 18: Project Files, World Game

Subseries 2, box 33, Folder 6: World Game History 2, General Correspondence 1967–69.

Palacin-Silva, Maria V., Antti Knutas, Maria Angela Ferrario, Jari Porras, Jouni Ikonen and Chandara Chea. 'The Role of Gamification in Participatory Environmental Sensing: A Study in the Wild'. In *Proceedings of the 2018 CHI Conference on Human Factors in Computing Systems (CHI'18)*, 2018, Association for Computing Machinery, New York, NY, USA, Paper 221: 1–13. doi:10.1145/3173574.3173795

Pask, Elizabeth. 'Today Has Been Going on for a Very Long Time'. *Systems Research* 10, no. 3 (1993): 143–147.

Pask, Gordon. 'Proposals for a Cybernetic Theatre'. 1964. Privately circulated monograph, Theatre Workshop and Systems Research, held at Pask Archive: http://www.pangaro.com/Pask-Archive/.

Pask, Gordon. *Conversation Theory: Applications in Education and Epistemology*. Amsterdam: Elsevier, 1976.

Pasztor, Andy. 'A Center to Study Implications of Technological Developments'. *The Colonial News*, Friday 13 March 1970: 6.

Perlmutter, Ellen. 'Poverty Can Be Erased, Sociologist Says'. *The Sun Bulletin*, 14 January 1976.

Pettifor, Ann. *The Case for a Green New Deal*. London: Verso, 2019.

Pickering, Andrew. *The Cybernetic Brain: Sketches of Another Future*. Chicago, IL: Chicago University Press, 2010.

Poundstone, William. *Prisoner's Dilemma*. New York: Anchor, 1993.

Raessens, Joost. *Homo Ludens 2.0: The Ludic Turn in Media Theory*. Utrecht: Utrecht University Press, 2012.

Ramage M. and K. Shipp, eds. *Systems Thinkers*. London: Springer, 2009.

Ray, Barbara. 'Panel to Investigate University Purposes'. *The Colonial News*, Friday 13 February 1970: 2.

Reffin Smith, Brian. 'Computer Art: Recent Trends'. *Computer Aided Design* 7, no. 4 (October 1975): 225–228.

Reichardt, Jasia, ed. *Cybernetics, Art and Ideas*. New York: Graphic Society, 1971.

Robbins, David, ed. *The Independent Group: Postwar Britain and the Aesthetics of Plenty*. Cambridge, MA: MIT Press, 1990.

Rooney-Varga, Juliette N., Florian Kapmeier, John D. Sterman, Andrew P. Jones, Michele Putko and Kenneth Rath. 'The Climate Action Simulation'. *Simulation & Gaming* 51, no. 2 (April 2020): 114–140.

Ryan, Patrick. 'Not Playing the Game'. *New Scientist* 57, no. 828 (11 January 1973): 96.

Sadler, Simon. 'The Dome and the Shack: The Dialectics of Hippie Enlightenment'. In *West of Eden: Communes and Utopia in Northern California*, edited by Iain Boal, Janferie Stone, Michael Watts and Cal Winslow. Oakland, CA: PM Press, 2012, 72–80.

Salgo, Peter. 'Innovational Projects Board'. *Pipe Dream*, Friday 23 April 1971.

Schlossberg, Edwin. World Game Diary, 1969. Stanford University Collections M1090, Series 18: Project Files, World Game Subseries 2, box 39, Folder 2, 12.

Schrape, Niklas. 'Gamification and Governmentality'. In *Rethinking Gamification*, edited by S. Fizek, M. Fuchs, P. Ruffino and N. Schrape. Lüneburg: Meson Press, 2014.

Schroeder, Richard. 'Woods Society—McHale and the Future'. *Pipe Dream*, Friday 8 March 1974, 13.

Scott, Felicity D. 'Fluid Geographies: Politics and the Revolution by Design'. In *New Views on R. Buckminster Fuller*, edited by Hsiao-Yun Chu and Roberto G. Trujillo. Stanford, CA: Stanford University Press, 2009, 160–175.

Scott, Felicity D. *Outlaw Territories: Environments of Insecurity/Architectures of Counterinsurgency*. New York: Zone Books, 2016.

Simon, Joe. 'Center's Move North Gives UB School "Window on the World"'. *The Spectrum*, Wednesday 19 September 1979: 3.

Smith, Giulia. 'Painting that Grows Back: Futures Past and the Ur-feminist Art of Magda Cordell McHale, 1955–1961'. *British Art Studies* 1 (2015). doi:10.17658/issn.2058-5462/issue-01/gsmith

Somit, Albert. Letter to Harold L. Cohen, dated 5 November 1979. University of Buffalo Special Collections, Center For Integrative Studies.

Stott, Timothy. 'Ludic Pedagogies at the College of Environmental Design, UC Berkeley, 1966 to 1972'. In *The Culture of Nature in the History of Design*, edited by Kjetil Fallan. London: Routledge, 2019, 58–71.

Sunwoo, Irene. 'Pedagogy's Progress: Alvin Boyarsky's International Institute of Design'. *Grey Room* 34 (Winter 2009): 28–57.

Sutcliffe, Alan. 'Patterns in Context'. In *White Heat Cold Logic: British Computer Art 1960–1980*, edited by Paul Brown, Charlie Gere, Nicholas Lambert and Catherine Mason. Cambridge, MA: MIT Press, 2009, 175–190.

Swade, Doron D. 'Two Cultures: Computer Art and the Science Museum'. In *White Heat Cold Logic: British Computer Art 1960–1980*, edited by Paul Brown, Charlie Gere, Nicholas Lambert and Catherine Mason. Cambridge, MA: MIT Press, 2009, 203–218.

Talbot, Linda. 'Meet the Friendly Robots'. *Hampstead and Highgate Express*, 26 July 1968.

Thom, René. 'At the Boundary of Man's Power: Play'. *SubStance* 8, no. 4 (1979): 11–19.

Thompson, Clive. 'QAnon Is Like a Game—A Most Dangerous Game'. *Wired*, 22 September 2020. https://www.wired.com/story/qanon-most-dangerous-multiplatform-game/

Tufte, Edward R. *Envisioning Information*. Cheshire, CT: Graphics Press, 1990.

Turner, Thomas. Week-by-week schedule for World Game Workshop at SIU, Summer 1970. Stanford University Collections M1090, Series 18: Project Files, World Game Subseries 2, box 105, Folder 8: Tom Turner Files, Workshop.

Turner, Thomas. 'World Game Facility'. Undated. Stanford University Collections M1090, Series 18: Project Files, World Game Subseries 2, box 105, Folder 9: Tom Turner Files, World Game Facility.

Turner, Thomas B. 'World Game: State-of-the-Art Report'. December 1969. Stanford University Collections M1090, Series 18: Project Files, World Game Subseries 2, box 25, folder 11 (1 of 3).

Tyrrell, Toby. *On Gaia: A Critical Investigation of the Relationship between Life and Earth*. Princeton, NJ: Princeton University Press, 2013.

Van Wijk, Jarke J. 'Unfolding the Earth: Myriahedral Projections'. *The Cartographic Journal* 45, no. 1 (2008): 32–42.

Vidler, Anthony. 'Whatever Happened to Ecology? John McHale and the Bucky Fuller Revival'. *Log* 13/14 (Fall 2008): 139–146.

Von Bertalanffy, Ludwig. 'An Outline of General System Theory'. *British Journal for the Philosophy of Science* I, no. 2 (August 1950): 134–165.

Von Bertalanffy, Ludwig. *General Systems Theory: Foundations, Development, Applications*. London: Allen Lane, 1971.

Ward, Barbara and René Dubos. *Only One Earth: The Care and Maintenance of a Small Planet*. New York: Norton, 1972.

Ward, Barbara. *Spaceship Earth*. New York: Columbia University Press, 1966.

Wasiuta, Mark. 'The Persistence of Informational Vision: World Game 1969, 2009'. *Journal of the Society of Architectural Historians* 68, no. 4 (2009): 590–593.

Watson, Greg. 'Education for a Design Science Revolution: Buckminster Fuller's World Game Workshop'. Unpublished presentation at Rhode Island School of Design, 5 November 2020.

Weibel, Peter. 'It Is Forbidden Not to Touch: Some Remarks on the (Forgotten Parts of the) History of Interactivity and Virtuality'. In *Media Art Histories*, edited by O. Grau. Cambridge, MA: The MIT Press, 2007, 21–41.

Wigley, Mark. *Buckminster Fuller Inc.: Architecture in the Age of Radio*. Zurich: Lars Müller, 2015.

Wigley, Mark. 'Recycling Recycling'. *Interstices: Journal of Architecture and Related Arts* 4 (2019): https://interstices.ac.nz/index.php/Interstices/article/view/589

Williams, R. John. 'World Futures'. *Critical Inquiry* 42 (Spring 2016): 473–546.

Winsey, Val. 'World Game Has Serious Ends'. *Pace Press*, Wednesday 11 March 1970: 10.

Winsey, Val. Letter to Buckminster Fuller, dated 7 May 1972. Stanford University Collections M1090, Series 18: Project Files, World Game Subseries 2, box 39, Folder 18: World Game History, Val Winsey (3 of 3).

Wood, Denis and John Fels. *The Natures of Maps: Cartographic Constructions of the Natural World*. Chicago, IL: University of Chicago Press, 2008.

World Economic Forum. *World Economic Forum—A Partner in Shaping History, The First 40 Years, 1971 to 2010*. Cologny: World Economic

Forum,   2009.   http://www3.weforum.org/docs/WEF_First40Years_Book_2010.pdf

World Game Corporate Program, Facilitator/Guide Script. Philadelphia, PA: World Game Institute, 1993.

World Game Workshop Teacher's Manual. Philadelphia, PA: World Game Institute, 1994.

Youngblood, Gene. 'Buckminster Fuller's World Game'. *Whole Earth Catalog,* March 1970.

Youngblood, Gene. 'Earth Nova'. *Los Angeles Free Press,* 3 April 1974.

Zimmerman, Eric. 'Position Statement: Manifesto for a Ludic Century'. In *The Gameful World: Approaches, Issues, Applications,* edited by Walz Steffen P. and Deterding Sebastian. Cambridge, MA: MIT Press, 2014, 19–22.

Index

Note: *Italic* page numbers refer to figures.